PPM

Practical Problems in Mathematics

FOR DRAFTING AND CAD

4TH EDITION

PPM

Practical Problems in Mathematics

FOR DRAFTING AND CAD

4TH EDITION

Dr. John C. Larkin

Professor Emeritus
Central Connecticut
State University

Dr. Concetta M. Duval

Duval & Associates Consulting

DELMAR
CENGAGE Learning·

Australia • Brazil • Japan • Korea • Mexico • Singapore • Spain • United Kingdom • United States

DELMAR
CENGAGE Learning·

Practical Problems in Mathematics for Drafting and CAD, Fourth Edition
John C. Larkin and Concetta M. Duval

Vice President, Careers & Computing:
Dave Garza

Director of Learning Solutions: Sandy Clark

Associate Acquisitions Editor: Kathryn Hall

Managing Editor: Larry Main

Product Manager: Mary Clyne

Editorial Assistant: Kaitlin Murphy

Vice President, Marketing: Jennifer Baker

Marketing Director: Deborah Yarnell

Associate Marketing Manager: Erica
Ropitsky

Senior Production Director: Wendy Troeger

Production Manager: Mark Bernard

Content Project Manager: Barbara LeFleur

Senior Art Director: David Arsenault

Media Editor: Deborah Bordeaux

For product information and technology assistance, contact us at
Cengage Learning Customer & Sales Support, 1-800-354-9706
For permission to use material from this text or product,
submit all requests online at **www.cengage.com/permissions**.
Further permissions questions can be e-mailed to
permissionrequest@cengage.com

Library of Congress Control Number: 2012932007

ISBN-13: 978-1-111-31680-8

ISBN-10: 1-111-31680-5

Delmar
5 Maxwell Drive
Clifton Park, NY 12065-2919
USA

Cengage Learning is a leading provider of customized learning solutions with office locations around the globe, including Singapore, the United Kingdom, Australia, Mexico, Brazil, and Japan. Locate your local office at: **international.cengage.com/region**

Cengage Learning products are represented in Canada by Nelson Education, Ltd.

To learn more about Delmar, visit **www.cengage.com/delmar**

Purchase any of our products at your local college store or at our preferred online store **www.cengagebrain.com**

Notice to the Reader
Publisher does not warrant or guarantee any of the products described herein or perform any independent analysis in connection with any of the product information contained herein. Publisher does not assume, and expressly disclaims, any obligation to obtain and include information other than that provided to it by the manufacturer. The reader is expressly warned to consider and adopt all safety precautions that might be indicated by the activities described herein and to avoid all potential hazards. By following the instructions contained herein, the reader willingly assumes all risks in connection with such instructions. The publisher makes no representations or warranties of any kind, including but not limited to, the warranties of fitness for particular purpose or merchantability, nor are any such representations implied with respect to the material set forth herein, and the publisher takes no responsibility with respect to such material. The publisher shall not be liable for any special, consequential, or exemplary damages resulting, in whole or part, from the readers' use of, or reliance upon, this material.

Printed in the United States of America
1 2 3 4 5 6 7 16 15 14 13 12

TABLE OF CONTENTS

SECTION 1   WHOLE NUMBERS

SECTION 2   FRACTIONS

SECTION 3   DECIMALS

SECTION 4 DECIMALS, FRACTIONS, AND PERCENTS

SECTION 5 GEOMETRY

SECTION 6 MEASUREMENT

SECTION 7 ALGEBRA

SECTION 8   GRAPHING

SECTION 9   APPLIED TRIGONOMETRY

SECTION 10   ESTIMATION AND TOLERANCE

PREFACE

The drafting profession has undergone dramatic changes since the 1980s. The single most significant change to occur is the widespread use of computer-aided drafting and computer-aided design systems (CAD) in all drafting fields. These systems increase the speed, accuracy, quality, and repeatability of the work that drafters do, and because of this, drafters have been able to increase their production using such systems.

Practical Problems in Mathematics for Drafting and CAD, 4th edition, has been revised to provide practical and real-life problem-solving experiences that drafters encounter in different drafting fields. This workbook is an excellent supplement to any vocational or basic drafting mathematics text. Students will find many opportunities to test their problem-solving capabilities, as well as to learn the symbolic language used within various drafting fields. In solving many drafting problems, students must analyze blueprints, sketches, and multi-view drawings and understand the role of the CAD drafter or operator in producing and interpreting drawings.

DELMAR'S PPM SERIES

This text is one of a series of workbooks designed to offer students practical problem-solving experiences within various occupations. The workbooks take a step-by-step approach to mastering basic mathematical skills. Each workbook includes relevant and easily understood problems in a specific vocational field. The workbooks are suitable for any student from middle or junior high school through high school and up to a two-year or perhaps even a four-year college level. To better help students, the appendix in this workbook contains descriptions of function keys found on most scientific calculators, tables of numerical and measurement equivalences, and a list of mathematical formulas and basic Geometric Dimensioning and Tolerancing (GD&T) symbols. In addition, a glossary of mathematical and drafting terms and answers to odd-numbered problems are also included. For more information about this series and a current list of titles, please visit www.CengageBrain.com.

SERIES FEATURES

The workbooks in Delmar's PPM series take a step-by-step approach to mastering essential mathematical skills. At the start of each unit, a brief introduction of the topic provides a basic explanation of the skills and concepts necessary to complete the problems in that unit. Examples are presented to help the learner review these principles. Problems in most units progress from basic skill-level examples to more complex problems that require critical thinking and analysis. As students progress through each unit, they will become more proficient at solving a wide variety of math and drafting problems.

THIS BOOK'S APPROACH

Practical Problems in Mathematics for Drafting and CAD, 4th edition, begins with updated reviews of the basic arithmetic operations with whole numbers, fractions, decimals, and percents, progresses through the basics of geometry, measurement, and algebra, and ends with sections on applied trigonometry, estimation and percent error, and tolerance. Topical sections are divided into short units to give teachers maximum flexibility in planning and to help students achieve maximum skill mastery. Instructors may choose to use this book as a stand-alone text or as a supplement to a theory-based text.

Within all types of examples and problems presented in this workbook, standardized procedures and conventional practices in the various fields of drafting have been adhered to. The problems in most units have been reorganized into one of three types:

- **Skill-Level** problems that deal with basic mathematical computations and serve as a review of basic mathematical skills with which future drafters should be comfortable.
- **Practical Problems** that provide relevant situations that are set mostly in the drafting world. Solving them requires applying basic math skills within realistic settings.
- **CAD Problems** that provide a wide variety of drawings that a drafter may confront in his or her profession. This set of problems provides the most comprehensive and challenging types of problems for students to solve. They require not only a firm grasp of underlying mathematical skills, but also the ability to apply analysis and reasoning skills.

NEW TO THIS EDITION

The fourth edition of *Practical Problems in Mathematics for Drafting and CAD* has been improved by expanding many of the mathematical explanations at the start of each unit, by providing additional stepped-out solutions to unique problems in most units, and by reorganizing the units within the 10 sections in a more coherent way. Major changes include the following:

- An expanded and reorganized section on GEOMETRY
- An expanded and reorganized section on ALGEBRA
- A new section dedicated to ESTIMATION AND TOLERANCE, including a brief introduction to GD&T.

The explanations and examples within each unit allow students many opportunities to review basic mathematical skills and concepts needed to solve problems in drafting with or without a calculator.

SUPPLEMENTS

The supplements package for this edition has been revised and expanded to include a new Instructor Companion Website and Applied Math CourseMate, a new online tool that can help students and teachers build lasting math skills.

Instructor Resources

The Instructor Companion Website provides the following support for teachers:

- Updated answers to all text problems
- Computerized test banks in ExamView® software
- PowerPoint® presentations
- An Image Gallery including all text figures

Applied Math CourseMate

Every text in Delmar's PPM series includes Applied Math CourseMate, Cengage Learning's online solution for building strong math skills. Students and instructors alike will benefit from the following CourseMate Resources:

- An interactive eBook, with highlighting, note taking, and search capabilities
- Interactive learning tools including:
  - Quizzes
  - Flashcards
  - PowerPoint slides
  - Skill-building games
  - and more!

Instructors will be able to use Applied Math CourseMate to access the Instructor Resources and other classroom management tools. Go to login.cengagebrain.com to access these resources.

ACKNOWLEDGMENTS

Delmar Cengage Learning and the authors would like to thank the following reviewers for their valuable suggestions and technical mathematical expertise:

Content Review

Joseph Connell, Santa Barbara City College
B. S. Sridhara, Middle Tennessee State University
Mark Freeman, Westech College School of Technology
Paul Nickels, Van Buren Technology Center

Technical Review

Linda Willey, Clifton Park, New York

ABOUT THE AUTHORS

Dr. John Larkin recently retired as Professor Emeritus of Technology Education from Central Connecticut State University, where he was an experienced drafting and CAD instructor. He is still active in numerous professional organizations. Dr. Larkin earned his doctorate from the University of Maryland.

Dr. Concetta Duval earned her doctorate from the University of Rochester, with a concentration in mathematics and mathematics education. She is currently a consultant, specializing in creating mathematics and science materials for K–12 students and teachers and published by a variety of textbook and computer education companies. Before working as a consultant, Dr. Duval was employed as a chemical engineer, a public high school math and science teacher, a school administrator, and director of mathematics at four private companies.

TO THE STUDENT

Drafting is concerned with drawings of objects. Drafters not only work "on the board" using standard drafting instruments, but also use computers and CAD software applications and printers to prepare working drawings. The drawings that drafters prepare must be accurate and need to include standardized labeling procedures. Other people must be able to read and interpret your work as a drafter or CAD operator.

Here is a summary of the skills you need to enjoy a successful career in drafting today.

- An understanding of the basics of drafting and design and underlying mathematical skills
- An ability to communicate ideas with freehand sketches
- An ability to use conventional drafting instruments quickly and clearly
- An ability to use computer-aided drafting and design systems[1]

Whether a drawing is created by hand or by a computer, a drawing is basically a way for you to communicate the design and manufacture of a part. All drawings consist of lines, and every line has a specific *dimension*. A dimension is a numerical value that is used to define the size, location, orientation, form, or other geometric property of a part. Some symbols may vary among different drawings and within different subfields, depending on the way a drawing is rendered. However, the important thing to remember is that whatever symbols you, as a drafter, may use, your drawings must be accurate and clear to the reader.

This workbook is divided into 10 sections. Each section deals with a major topic in mathematics and how it can be used to solve drafting and CAD problems. Mathematical topics include a review of basic operations in arithmetic, algebra, geometry, and trigonometry. As described earlier, this workbook edition contains three types of problem sets in nearly each unit: skill-level problems, practical problems, and CAD problems. Each problem set provides you with many opportunities in which to apply your math and drafting skills.

The field of drafting is a broad one and includes several subfields, such as mechanical engineering, civil engineering, and architectural engineering. Each subfield may have its own set of conventions and graphic symbols. Therefore, you will see a wide variety of CAD drawings in this workbook as examples from various subfields. Each drawing has been chosen to give you a general idea of the many types of drawings and symbols that arise in this field.

The following section is a brief introduction to a few of the basic conventions and symbols used in most drafting diagrams Study these descriptions so that they will help you interpret and solve the many CAD problems in this workbook.

---

[1]P. R. Wallach *Fundamentals of Modern Drafting.* (Thomson, Delmar Learning: Clifton Park, NY, 2003), 2.

LINES

The lines used in a CAD drawing to draw an object are always **thick** lines. These lines define the object and all of its parts. The auxiliary lines used to define dimensions or particular features of an object are always **thin** lines.

Dimension lines: These thin lines show the linear distance between two points. Arrowheads at the ends of a dimension line show the beginning and end of the intended measurement. These lines may be on the outside of an object if the measurement is small.	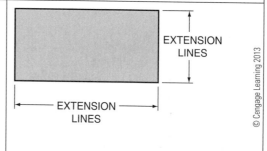
Extension lines: These are thin lines that extend from the ends of an object and that indicate the beginning and end of a measured distance or a distance to be measured. In some drawings, as in the architectural drawing shown here, distances between extension lines are indicated by diagonal slashes(/).	
Leader lines: These lines identify a given part or dimension. In many cases, leaders have an arrow head at one end that points to a specific part of a drawing.	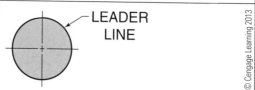

Center lines: These are thin, broken lines made up of long and short dashes. They may indicate the center of a circle (A) or a part of a circle (B). They also are used to indicate that an object is symmetrical (C).	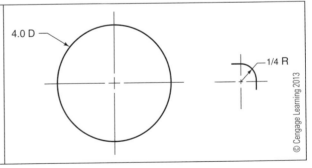
Hidden edges: These are thin, dashed lines that indicate a part of an object that is hidden from view.	

CIRCLES

A circle in drafting is used to represent a cylindrical structure or part, or a hole that extends all the way through an object, unless its depth is specified.

The arrowhead at the end of a leader line stops at a point on the circle. The other end of the leader line may display the dimension of a radius (R or RAD), a diameter (D or DIA) of the circle, or an arc (part of the circle).	

If holes are equally spaced around a circle, the exact location of the first hole is shown. The diameter, the size of the holes, the number of holes, and a notation "EQUALLY SPACED" are also indicated.

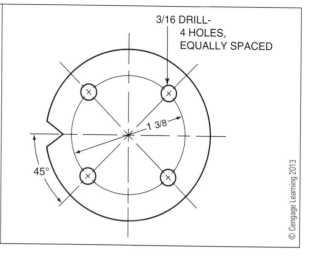

© Cengage Learning 2013

UNITS

Measurement units, such as inches, feet, centimeters, and so on are not always included within a drawing. A special area on a drawing is set aside for information about the units used to create the drawing of the object. However, within any one drawing, the units used for similar parts of objects are always the same. These units are from either the U.S. customary system or the metric system.

COMMON ABBREVIATIONS

D, DIA, $\varnothing$	diameter of a circle
R, RAD	radius of a circle
TYP	typical
PLCS	places
BC	bolt circle[2]
DIST	distance
OC	on center
DIM	dimension

---

[2]A bolt circle is an imaginary circle determined by the position of the bolts. The center of each bolt lies on the bolt circle itself.

EXAMPLE:

Some of the traditional conventions and symbols can be seen in the CAD drawing below. Notice that the **thick** lines define the shape of the object and the parts (holes) within it. The **thin** lines indicate properties of the object and its parts. Study this figure to see how the dimensions on this CAD drawing of an object have been specified.

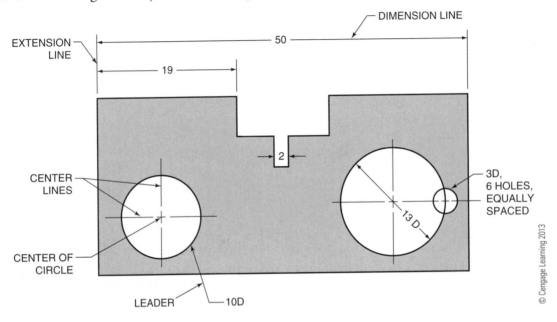

- The dimension line at the top of the figure represents the length of the object, 50.
- The three extension lines indicate the lengths of the two dimension lines, 19 and 50.
- Smaller dimension lines indicate the length of dimension 2.
- Leader lines on the left point to the center lines, the center of the circle, and the length of its diameter.
- The smaller circle on the left has a diameter of 10 (10D), and the larger circle on the right has a diameter of 13 (13D).
- There are six holes to be equally spaced around the circle on the right. The diameter of each of these holes is 3 (3D).

SECTION

Whole Numbers

UNIT 1

Addition

Basic Principles

The set of whole numbers consists of the number 0 and all of the counting numbers: {0, 1, 2, 3, …}. A whole number has no fractional or decimal part.

A place-value chart shows the value of each digit in a whole number, starting with the ones place on the far right, the tens place, the hundreds place, and so on.

millions	hundred thousands	ten thousands	thousands	hundreds	tens	ones

© Cengage Learning 2013

To add two or more whole numbers, line up the digits in each number according to its place value.

EXAMPLE: Find the sum. $19 + 7 + 76 + 113 + 258$

STEP 1:

Start at the far right and add the digits in the ones place first. The sum is 33 ones.

hundreds	tens	ones
	1	9
		7
	7	6
1	1	3
+2	5	8
		33

© Cengage Learning 2013

STEP 2:

Thirty three in the ones place is greater than 9, so regroup the tens digit into the tens place of the chart. Then add the tens. The sum is 17 hundreds.

hundreds	tens	ones
	3	
	1	9
		7
	7	6
1	1	3
+2	5	8
	17	3

© Cengage Learning 2013

STEP 3:

Seventeen hundreds is greater than 9 hundreds, so regroup the hundreds digit into the hundreds place of the chart. Add the hundreds. The sum is 4 hundreds.

hundreds	tens	ones
1	3	
	1	9
		7
	7	6
1	1	3
+2	5	8
4	7	3

© Cengage Learning 2013

 19 ＋ 7 ＋ 76 ＋ 113 ＋ 258

Skill Problems

Add the following quantities.

NOTE: When adding numbers with units of measure, all units must be the same.

1.   31 inches
    60 inches
    46 inches
  + 83 inches

2.   315 feet
    90 feet
    216 feet
  + 451 feet

3.   305 millimeters
    167 millimeters
    236 millimeters
  + 133 millimeters

Practical Problems

4. A civil drafter uses a drafter's scale to measure these lengths: 6 inches, 17 inches, 34 inches, 63 inches, 26 inches, and 9 inches. What is the total length of the line in inches?

5. A drafting department supervisor orders several types of drafting pencils for the department. The order is for 24 HB pencils, 16 H pencils, 27 2H pencils, 31 3H pencils, 48 4H pencils, 36 F pencils, 30 2B pencils, and 9 6H pencils. What is the total number of pencils?

6. To determine a drafting team's use of time, a supervisor makes this chart.

TIME RECORD
(in hours)

DRAFTER	PROJECT					HOURS (a)
	A	B	C	D	E	
Frank	6	5	3	4	1	
Carol	3	4	7	0	5	
Bob	1	4	2	6	3	
Tanya	3	6	4	2	5	
Jim	5	1	2	4	3	
TOTAL HOURS (b)						

   a. Calculate the total number of hours each drafter works and enter each value in the table above.

   b. Calculate the total number of hours the whole team works and enter the value in the table above.

7. An office clerk reports the total number of blueprints made each week. On Monday, there were 26 reports; on Tuesday, 21; on Wednesday, 21; on Thursday, 37; and on Friday, 17. How many blueprints were made this week?   _____

8. A map drafter is taking inventory of his drafting equipment. There are 15 pencils, 6 erasers, 3 T-squares, 5 triangles, 2 protractors, 7 inking pens, and 4 irregular curves. How many pieces of equipment are on hand?   _____

9. An architectural drafter uses many different dimensions to block in the views of an object. The individual measurements of the lengths of the object are 4′, 7′, 9′, 3′, 6′, and 8′. What is the total length in feet of the object?   _____

10. A detail drafter turns in his time card showing the time he spent on six different jobs. The six times, expressed in minutes, are 236, 757, 520, 418, 357, and 132. What is the total time, in minutes, that he spent on these jobs?   _____

11. An assembly drawing contains six sheets of detail drawings. The numbers of details on the six sheets are 7, 11, 9, 5, 8, and 7 respectively. What is the total number of details?

_____

12. A certain computer keyboard contains 28 letters, 10 numerals, 12 function keys, 9 numeric pad keys, 11 punctuation keys, and 28 miscellaneous keys. What is the total number of keys on this keyboard?

_____

13. A CAD input tablet menu contains 26 crosshatching symbols, 39 architectural symbols, 18 dimensioning commands, 15 editing commands, and 9 inquiry commands. What is the sum of the commands and symbols contained in these five categories on the tablet menu?

_____

14. An architectural drafter needs to calculate the total number of doors of different types on drawings for a new project. The drawings contain the following types of doors: 47 sliding doors, 35 accordion doors, 17 bi-fold doors, 15 french doors, 12 dutch doors, and 11 pocket doors. What is the total number of doors for this project?

_____

15. A CAD drafter obtains a directory of drawing files from an external hard drive. The files contain the following number of bytes: 239,675; 81,300; 347,000; 5,953; 67,307; 176,354; and 1,739 bytes.

   a. Find the sum of the three smallest files.   a. _____

   b. Find the sum of the three largest files.   b. _____

   c. Find the total number of bytes for all drawing files.   c. _____

16. A CAD drafter obtains a directory of all part and pattern files used with a new project. The directory of files appears below. Determine how many bytes are contained in all of the parts (*.PRT) and pattern (*.PTN) files.

BASE.PRT	46,971	HEXNUT.PTN	1,124
BLOCK.PRT	31,921	LEVER.PRT	3,112
CAPSCREW.PTN	2,874	PUNCH.PRT	5,362
CAPSCR1.PTN	2,372	SHAFT.PRT	1,778
CLAMP.PRT	7,928	WASHER.PTN	876
GASKET.PTN	4,521		

Part (*.PRT) files _____

Pattern (*.PTN) files _____

17. A CAD drawing contains many features (lines, arcs, circles, and so on). What is the total number of features in a CAD drawing that contains 107 object lines, 21 center lines, 17 arcs, 9 hidden lines, and 6 circles? _____

18. A CAD-operators user group meets once a month. Last month, the number of operators from nine area towns and cities were 11, 17, 8, 21, 13, 7, 12, 16, and 15. How many operators were at the meeting last month? _____

CAD Problems

19. What is the overall unit length of this link? _____

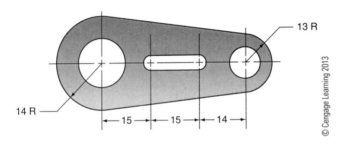

20. What is the overall unit lengths of dimensions **A** and **B** on this foundation plan?

A _____

B _____

21. The *perimeter* of a figure is the total distance around the outside of the figure. What is the perimeter of this shim? (A shim is a thin and often tapered or wedged piece of material.)

_____

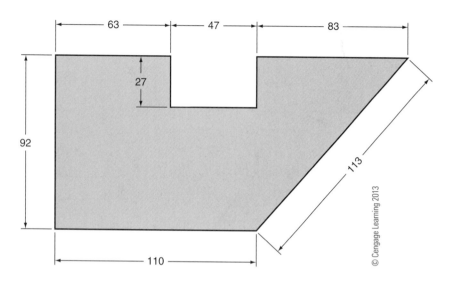

22. a. What is the length of the support block shown below?
    b. What is the height of this support block?

a. _____

b. _____

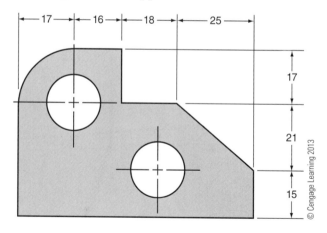

23. What is the perimeter of this plate?   _____

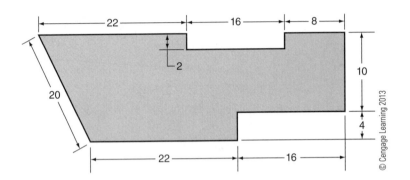

24. Use the CAD drawing below to calculate dimensions **A** and **B**.

A _____

B _____

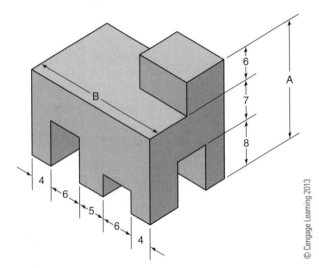

25. Use this CAD drawing of a gasket to calculate dimensions **A** and **B**.

A _____

B _____

26. The CAD drawing below shows the dimensions of a shaft. Find the lengths of dimensions **A, B, C,** and **D**.

A _____

B _____

C _____

D _____

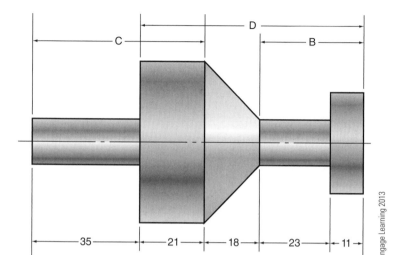

27. The CAD drawing below shows dimensions of a gauge. Calculate dimension **G**.

_____

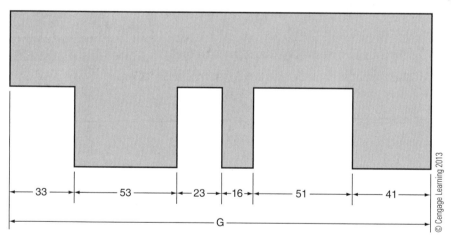

28. The drawing below represents the plot plan for the proposed site of a new building.

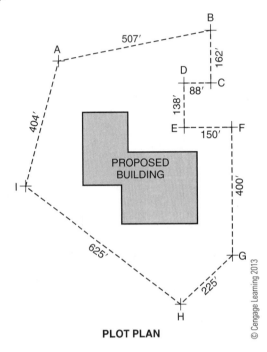

PLOT PLAN

a. Determine the perimeter of the plot in feet.

b. Express the sum of the four shortest boundary lines in feet.

c. Express the sum of the three longest boundary lines in feet.

a. _____

b. _____

c. _____

29. The CAD drawing below shows a symmetrical gasket.

    a. Calculate the perimeter of the gasket.

    b. Find the sum of the perimeters of the five internal features of the gasket.

a. _____

b. _____

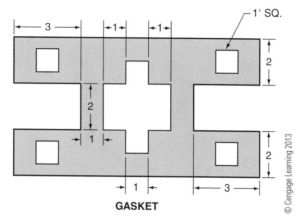

GASKET

30. Use the CAD drawing of the front elevation of this house to calculate its overall height and length in feet and inches. Include the measure of the roof overhang in the answer.

Height _____

Length _____

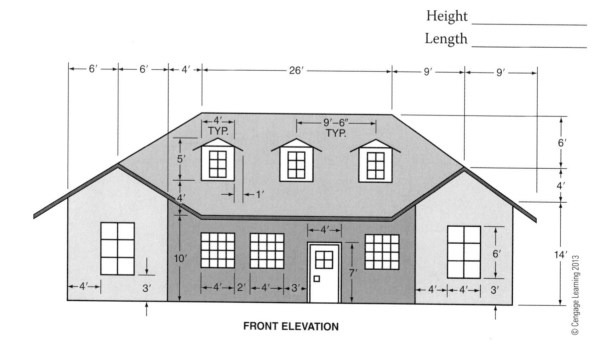

FRONT ELEVATION

31. Use the CAD drawing below to determine dimensions **A** through **E**.

A _____

B _____

C _____

D _____

E _____

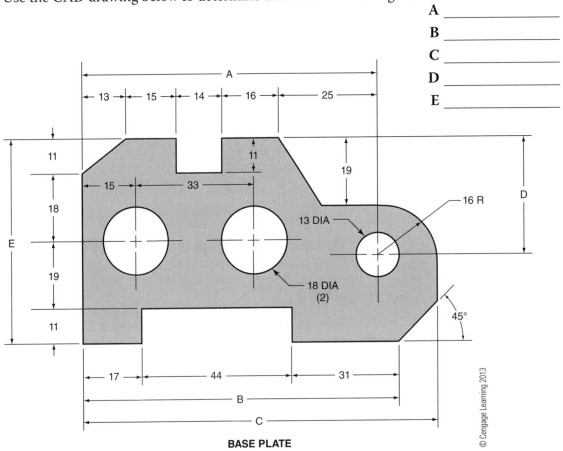

BASE PLATE

UNIT 2

Subtraction

Basic Principles

Subtraction is the process of finding the *difference* between two numbers. To subtract two whole numbers, line up the digits in each number in the correct columns of the place-value chart. Then start by subtracting the digits in the ones place.

EXAMPLE: Find the difference. 867 – 523

$$\begin{array}{r} 867 \\ -\ 523 \\ \hline 344 \end{array}$$

The difference is 344.

EXAMPLE: 386 – 57

If the digit in a place being subtracted is greater than the digit above it, regroup the number in the place to the left of that place.

13

STEP 1:

Seven in the ones place is greater than 6 in the ones place above it. So regroup 8 tens as 7 tens plus 1 ten.

$$80 = 70 + 10$$

Add 10 ones to 6 ones in the ones place: $16 - 7 = 9$.

STEP 2:

Now subtract in the tens place: 7 tens minus 5 tens is 2 tens.

STEP 3:

Finally, subtract in the hundreds place: 3 hundreds minus zero hundreds is 3 hundreds.

hundreds	tens	ones
	7	16
3	8̸	6̸
	−5	7
		9

© Cengage Learning 2013

hundreds	tens	ones
	7	16
3	8̸	6̸
	−5	7
	2	9

© Cengage Learning 2013

hundreds	tens	ones
	7	16
3	8̸	6̸
	−5	7
3	2	9

© Cengage Learning 2013

The difference is 329.

 386 ⊟ 57 ⊜

Skill Problems

Subtract. Include units in your answers where appropriate.

1. 108 feet
    − 16 feet

2. 102 inches
    − 37 inches

3. 457 millimeters
    − 206 millimeters

4. 1,476 pounds
    − 808 pounds

5. 325 yards
    − 116 yards

6. 263 inches
    − 79 inches

7. 585 feet
    − 183 feet

8. 4,153 millimeters
    − 1,276 millimeters

Practical Problems

9. A certain drafter works 226 days in one year (365 days). How many days is the drafter not working?

10. The GROUP command in a CAD program links entities in a drawing so they can be selected as a complete unit. A CAD drawing contains 1,465 entities in one group. Seventeen entities are deleted from the front view, 21 from the top view, 37 from the side view, and 43 from the pictorial view. How many entities are left in the group after these specific entities are deleted?

11. During one month, 220 working drawings are submitted to a company's checking team to check for accuracy. Only 187 drawings are checked and returned to the drafters. How many drawings were not returned?

12. An assembly drawing has 83 parts. The first drafter details 14 parts. The second drafter details 21 parts. The third drafter details 17 parts. How many parts are not yet drawn?

13. A CAD drafter's time limit for a specific job is 525 hours. The drafter recorded the following number of hours she worked each day on her time sheet: 40 hr., 36 hr., 50 hr., 48 hr., 40 hr., 42 hr., 38 hr., and 44 hr. How many more hours does she have left to finish the job?

14. A large company has 300 drafters. During the year, 45 were laid off, 23 retired, 37 left the company, and 8 were promoted out of the department. How many additional drafters need to be hired to maintain a staff of 250 drafters in the company?

15. In a certification program, there are 115 drafting students in four classes. One class has 28 students, the second class has 26 students, and the third class has 27 students. How many students are in the fourth class?

16. A drafter needs to find the length of one of the nine dimensions on a drawing. The overall length of the line is 284 mm. The lengths of the eight other parts on the line measure 17 mm, 35 mm, 67 mm, 27 mm, 52 mm, 26 mm, 11 mm, and 23 mm. Determine the length in millimeters of the last part of the line.

17. A CAD drawing of a large building contains a total of 915 symbols, including 110 windows, 72 doors, 33 appliances, 215 lighting fixtures, 365 electrical switches, and 89 mechanical equipment symbols. How many other symbols are on the drawing?   _____

18. A large assembly drawing contains 96 different parts that need to be detailed. A team of CAD drafters has detailed the following numbers of parts: 7, 5, 11, 15, 9, 13, and 8. How many parts still need to be detailed?   _____

19. A CAD operator has 256 levels available that he can use for drawing and plotting. He is currently using 126 levels. How many levels can he use if the following levels are turned off?   _____

    Construction Details 21   East Elevations 3
    Foundation Details 5   West Elevations 3
    Door Schedules 7   South Elevations 3

CAD Problems

20. What is the threaded length of this round-head machine screw?   _____

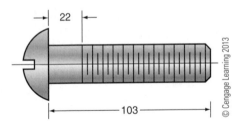

© Cengage Learning 2013

21. Calculate the overall length of this T-square.   _____

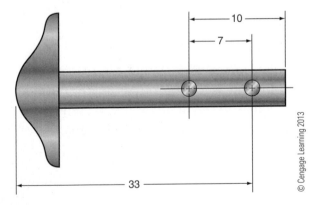

© Cengage Learning 2013

22. Determine lengths **A, B,** and **C** on this template.

A _____
B _____
C _____

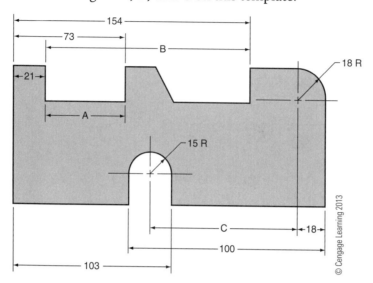

23. Use the top and front views of this figure to calculate distances **A, B,** and **C**.

A _____
B _____
C _____

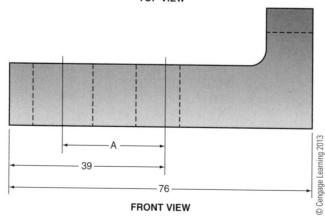

TOP VIEW

FRONT VIEW

24. The plot plan below shows property lines and setbacks. What is the difference in feet between the perimeter of the property and the perimeter of the setback, if the lengths of the property lines measure 210 ft. × 125 ft.?

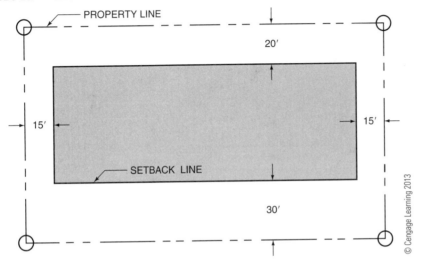

PROPERTY LINE

20'

15'   15'

SETBACK LINE

30'

© Cengage Learning 2013

25. Use the CAD drawing below to find the difference in feet between the perimeter of the property line and the perimeter of the house.

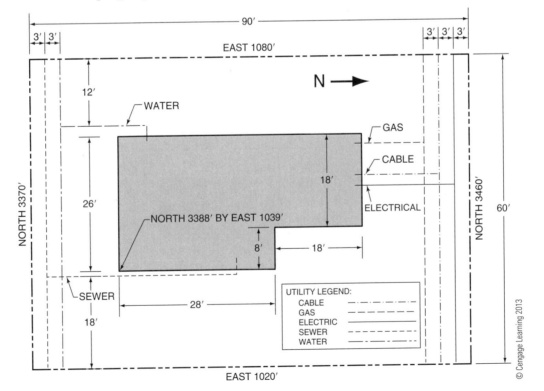

90'

3' 3'   3' 3' 3'

EAST 1080'

N →

12'

WATER   GAS

CABLE

18'

NORTH 3370'   26'   ELECTRICAL   NORTH 3460'   60'

NORTH 3388' BY EAST 1039'

8'   ← 18' →

SEWER   28'

18'

UTILITY LEGEND:
CABLE
GAS
ELECTRIC
SEWER
WATER

EAST 1020'

© Cengage Learning 2013

26. A CAD drafter needs to know the overall height and width for place-
ment purposes. Use the CAD drawing below to find the overall height
and width of this drawing.

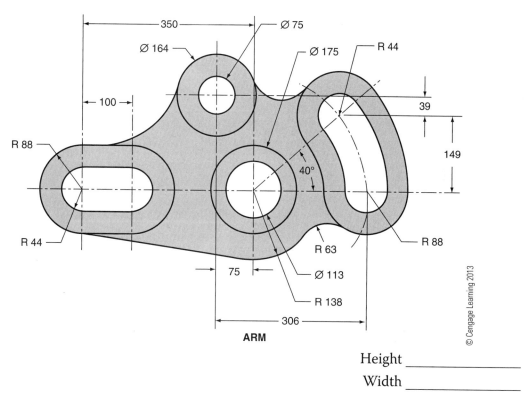

ARM

Height _____

Width _____

27. What is the length of radius **A** in the CAD drawing of this rocker arm? _____

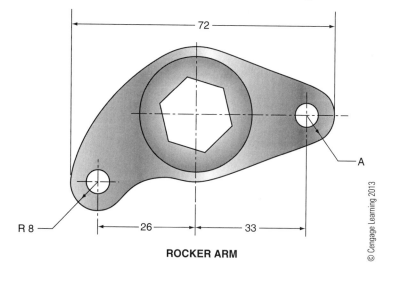

ROCKER ARM

© Cengage Learning 2013

28. What are dimensions **A, B,** and **C** in the CAD drawings below?

A _____

B _____

C _____

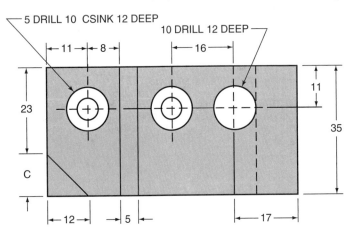

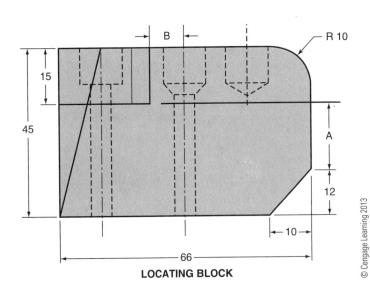

LOCATING BLOCK

© Cengage Learning 2013

29. What is the difference in feet between the sum of the perimeters of the house, garage, and driveway and the perimeter of the property lines in the CAD drawing below?

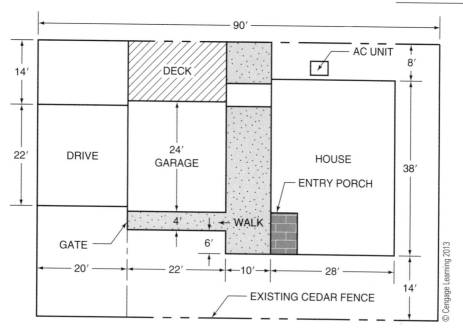

UNIT 3

Multiplication

Basic Principles

Two (or more) numbers being multiplied are called *factors.* The result is called the *product.*

$$6 \times 4 = 24 \leftarrow \text{product}$$

factors

One way to think about multiplying two whole numbers is called *repeated addition.*

EXAMPLE: $6 \times 4 = 4 + 4 + 4 + 4 + 4 + 4$

The product of 6 and 4 is equal to the sum of six 4s. Both the product and the sum equal 24.

NOTE: A unit-less factor in a multiplication problem that involves dimensions is called a *scalar.* Scalars are positive numbers that are frequently used in drafting to enlarge (scale up) or reduce (scale down) dimensions in a drawing. The product of a number and a scalar has the same unit as the unit of the dimension being scaled.

EXAMPLE: To scale up a dimension of 45 mm, multiply 45 mm by a scale factor of 5.

$$5 \times 45 \text{ mm} = 225 \text{ mm}$$

Multiplying numbers that have multiple digits requires regrouping.

EXAMPLE: Find the product. 4×375

STEP 1: Write the factors vertically with the smaller number at the bottom.

$$\begin{array}{r} 375 \\ \times\ 4 \\ \hline \end{array}$$

STEP 2:
- Multiply the digits in the ones place first.
  $4 \times 5 = 20$.
- Write a 0 in the ones place of the product and add 2 tens (20) to the tens place.

$$\begin{array}{r} {}^{+2} \\ 375 \\ \times\ 4 \\ \hline 0 \end{array}$$

STEP 3:
- Next multiply 7 tens (70) by 4.
  $4 \times 70 = 280$.
- Add the 2 tens (20, from STEP 2) to 280. The sum is 300.
- Write a 0 in the tens place of the product and add 3 hundreds to the hundreds place.

$$\begin{array}{r} {}^{+3+2} \\ 375 \\ \times\ 4 \\ \hline 00 \end{array}$$

STEP 4:
- Multiply 3 hundred (300) by 4.
  $4 \times 300 = 1200$.
- Add 3 hundreds (300 from STEP 3) to 1200. The sum is 1500.
- Write a 5 in the hundreds place of the product and a 1 in the thousands place of the product.

$$\begin{array}{r} {}^{+3+2} \\ 375 \\ \times\ 4 \\ \hline 1500 \end{array}$$

The product of 4 and 375 is 1500.

 4 ⏹×⏹ 375 ⏹=⏹

Skill Problems

Multiply the following dimensions by each scalar. Include the proper units in the answers.

1. $\begin{array}{r} 74 \text{ cm} \\ \times\ 9 \\ \hline \end{array}$

2. $\begin{array}{r} 107 \text{ lb.} \\ \times\ 8 \\ \hline \end{array}$

3. $\begin{array}{r} 345 \text{ in.} \\ \times\ 15 \\ \hline \end{array}$

4. $\begin{array}{r} 133 \text{ hr.} \\ \times\ 18 \\ \hline \end{array}$

5. $\begin{array}{r} 17 \text{ ft.} \\ \times\ 13 \\ \hline \end{array}$

6. $\begin{array}{r} 65 \text{ in.} \\ \times\ 43 \\ \hline \end{array}$

7. $\begin{array}{r} 36 \text{ yd.} \\ \times\ 24 \\ \hline \end{array}$

8. $\begin{array}{r} 29 \text{ mm} \\ \times\ 48 \\ \hline \end{array}$

Practical Problems

9. A detail drafter measures off eight line segments along a line. Each
    segment is 3 inches long. How long, in inches, is the line?   _____

10. A CAD drafter works on an assembly drawing 8 hours each day.
    The drafter estimates that it will take 8 more work-days to complete
    the drawing. Find the total number of hours worked on this job if the
    drafter's estimate is correct.   _____

11. A drafting department supervisor orders 17 reams of 18" × 24" paper.
    Each ream weighs 7 pounds. What is the total weight, in pounds, of
    this paper?   _____

12. A *checker* is a person who carefully reviews drawings made by
    drafters to see if they are correct. If a checker examines an average
    of 9 drawings a day, how many drawings are checked for each of the
    following numbers of days?

    a. 5 days   a. _____

    b. 20 days   b. _____

    c. 30 days   c. _____

13. A CAD operator needs to calculate the total cost of the wall switches
    shown on a first-floor plan. There are 17 single-pole, single-throw
    switches at $1.79 each; 19 single-pole, double-throw switches at
    $2.17 each; and 9 double-pole, double-throw switches at $3.29 each.

    a. What is the total cost of switches for the first floor?   a. _____

    b. Using the total cost as an average cost, what is the cost of the
       switches for the remaining 12 floors?   b. _____

    c. What is the cost of all the switches in the building?   c. _____

14. A byte contains 8 bits of information. How many bits are contained in
    a drawing file that has 1,973 bytes?   _____

15. A CAD drafter must determine how many symbols are contained on a
    recently completed project. Levels 1, 3, and 7 each have seven symbols;
    levels 2, 5, and 9 each have 11 symbols; levels 4, 8, and 16 each contain

12 symbols; levels; 6, 10, and 15 each have eight symbols; and levels 11, 12, 13, and 14 each have nine symbols. Find the total number of symbols used for this project.

_____

16. A CAD drafter must calculate the total number of doors on a set of plans for a large apartment building. The drafter determines the following total number of doors: 3 floors contain 42 doors each, 9 floors contain 35 doors each, and 15 floors contain 28 doors each. How many doors are contained in the set of plans for the apartment building?

_____

17. Using the same set of drawings as in Question 16, the CAD drafter must also calculate the total number of windows. He finds the following window totals: 16 floors contain 26 windows each, 7 floors contain 28 windows each, and 3 floors contain 23 windows each. Find the total number of windows contained in the set of plans for the apartment building.

_____

CAD Problems

18. Find the total length in millimeters, of artgum needed to make 7 erasers. _____

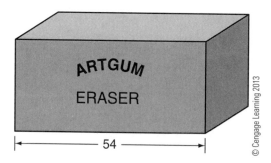

ARTGUM
ERASER

54

© Cengage Learning 2013

19. Through a new manufacturing process, bar stock originally measuring 157 mm can be shortened by 35 mm. If the new process is used, find the total length, in millimeters, of stock needed for 25 pieces.

_____

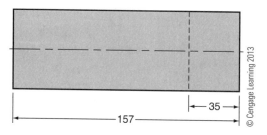

35

157

© Cengage Learning 2013

20. A hexagon is a two-dimensional figure that has six sides. The end portion of this piece of brass stock is a hexagon whose sides are all equal. Find the perimeter of the end portion of this stock.

_____

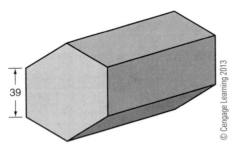

39

© Cengage Learning 2013

21. All holes in this drilled plate are equally spaced. Determine the total length from the center of hole **A** to the center of hole **H**.

_____

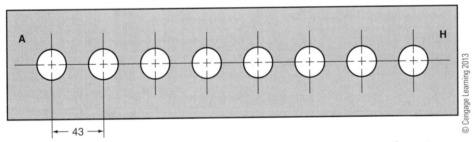

A

H

43

© Cengage Learning 2013

22. There are 100 centimeters in 1 meter. How many centimeters long is this 23-meter strap?

_____

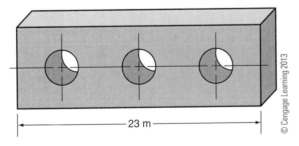

23 m

© Cengage Learning 2013

23. There are 10 decimeters in 1 meter. How many decimeters are contained in the total length of this 17-meter connecting link? _____

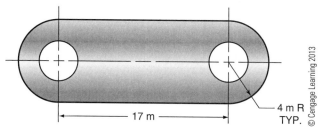

17 m

4 m R
TYP. © Cengage Learning 2013

24. A CAD drafter must locate seven 1-in. holes at 35° intervals around a circle. Determine the number of degrees from the center of hole **1** to the center of hole **7**. _____

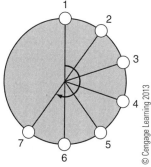

© Cengage Learning 2013

25. All sections of this sheet metal development, except the tab, have equal lengths. Calculate the overall length of this sheet, including its tab. _____

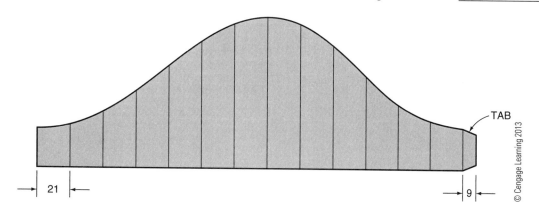

TAB

21   9

© Cengage Learning 2013

26. CAD drafters use the SCALE command to increase or decrease the size of objects. Determine the new height and width of this CAD drawing if it is enlarged by a factor of 4.

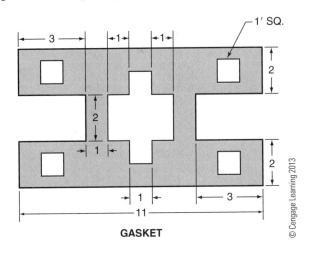

GASKET

Height _____

Width _____

27. Use this CAD drawing of a rocker arm and calculate its overall width and the lengths of dimensions **A** through **E** if the arm is scaled by a factor of 6.

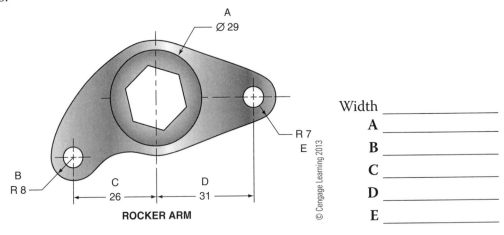

ROCKER ARM

Width _____

A _____

B _____

C _____

D _____

E _____

28. The CAD drawing below is to be scaled up by a factor of 3. Determine
     the perimeters in feet of the new drive, the garage, and the house.

Drive _____

Garage _____

House _____

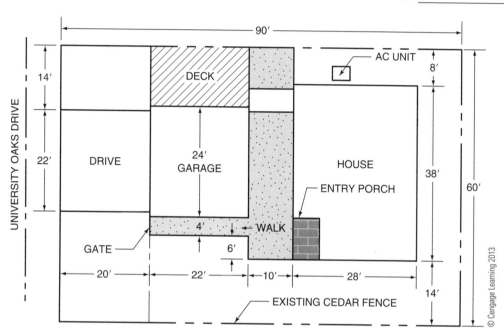

29. The CAD drawings for the locating block below is to be scaled by a factor of 9. Find the new sizes for dimensions **A** through **E** after the scaling operation has been completed.

A _____

B _____

C _____

D _____

E _____

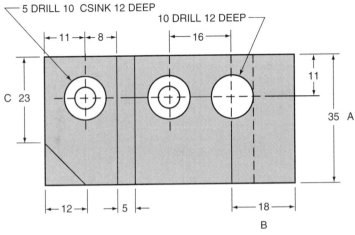

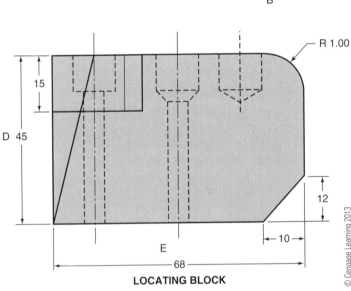

LOCATING BLOCK

30. The CAD drawing of this arm is to be scaled up by a factor of 6.
Calculate the new lengths of dimensions **A** through **E**.

A _____

B _____

C _____

D _____

E _____

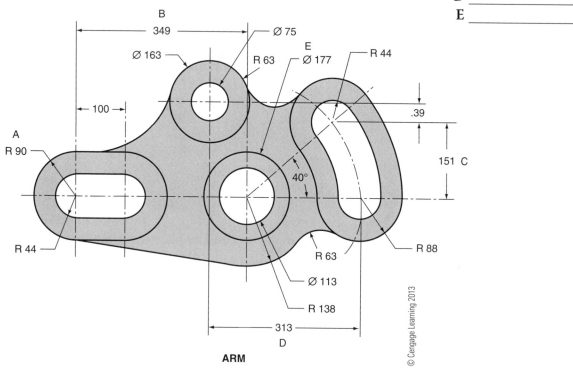

ARM

© Cengage Learning 2013

UNIT 4

Division

Basic Principles

Division is the opposite operation of multiplication. The number to be divided is called the *dividend.* The number used to indicate the number of times the dividend is to be divided is called the *divisor,* and the answer is called the *quotient.* If a number does not divide evenly into a dividend, there is a *remainder.* The remainder is always less than the divisor and can be expressed as a fraction whose numerator is the remainder and whose denominator is the divisor.

$$\text{divisor}\overline{)\text{dividend}}^{\,quotient\ +\ remainder}$$

Dividing numbers involves several steps involving multiplication and subtraction.

EXAMPLE: Calculate $1310 \div 15$.

STEP 1:   It's usually a good idea to start by estimating the quotient first, using trial and error. A reasonable estimate for this quotient is a number between 80 and 100.

$80 \times 15 = 1200$
$? \times 15 = 1310$
$100 \times 15 = 1500$

Because $1200 < 1500$ and the dividend, 1310, is closer to 1200, the quotient is between 80 and 100, closer to 80.

STEP 2:
- Set up the long division problem.
- Using the estimate, the tens digit of the quotient is 8. So write 8 above the tens digit in the dividend. (8 tens = 80)
- Multiply 8 tens (80) and 15 and write the product 120 tens under the dividend.
- Subtract 120 from 131.

$$15\overline{)1310}$$

$$\begin{array}{r} 8 \\ 15\overline{)1310} \end{array}$$

$$\begin{array}{r} 8 \\ 15\overline{)1310} \\ -120 \\ \hline 11 \end{array}$$

STEP 3:
- Bring down the 0 in the ones place of the dividend to write 110. That is the new dividend.
- Divide 15 into 110: $7 \times 15 = 105$.
- Write 7 in the ones place of the quotient.
- Subtract 105 from 110. The remainder is 5.

The answer is 87⁵⁄15, which equals 87⅓.

$$\begin{array}{r} 87 \\ 15\overline{)1310} \\ -120 \\ \hline 110 \\ -105 \\ \hline 5 \end{array}$$

📇 1310 ÷ 15 =

NOTE: The result using a calculator may be a decimal that is equivalent to 87⅓, which is approximately equal to 87.33.

Skill Problems

Divide. Include units in your answers as appropriate.

1. $6\overline{)448}$

2. 801 cm ÷ 9

3. $13\overline{)819}$

4. 1404 in ÷ 27

5. $166\overline{)4408}$

6. 459 mm ÷ 17

Practical Problems

7. An architectural drafter lays out (measures off) a line 72 inches long. The line is divided into 12-inch lengths. How many sections are obtained from this line?

    _____

8. How many 9-inch line segments lie along a straight line that is 144 inches long?

    _____

9. Six CAD drafters spend a total of 558 hours working on a special job. If each one works the same number of hours, how many hours does each drafter work?

    _____

10. The total weight of 40 drafting boards is 200 pounds. If the boards each have the same weight, how many pounds does each board weigh? _____

11. Three dozen erasing shields cost 12 dollars. What is the cost of 1 dozen erasing shields?

    _____

12. How many 4-inch erasing shields can be cut from a strip of sheet metal 128 inches long? (No allowance is made for waste.)

    _____

13. A drafter can make 5,040 triangles in 7 hours. Working at the same rate, how many can he make in each of the following numbers of hours?

    a. one hour   a. _____

    b. one minute   b. _____

14. A package of 2H drafting pencils contains 12 pencils. How many packages can be made from a box containing 900 pencils?

    _____

15. Determine how many bytes are contained in a drawing file containing 38,312 bits of information. (1 byte = 8 bits)

    _____

16. Five drawing files contain the following numbers of bytes: 12,337; 21,142; 10,615; 26,457; and 19,579. Calculate the average size of the files in bytes. (The average of a set of numbers is the sum of the numbers divided by the number of numbers in the set.)

    _____

17. A CAD drafter uses the SEGMENT command to divide each of two lines into equal segments.

   a. What are the lengths of the segments if line A is 153 mm long and is divided into 17 equal segments?   a. _____

   b. What are the lengths of the segments if line **B** is 143 mm long and is divided into 13 equal segments?   b. _____

18. A CAD operator wishes to set up the 256 levels used for drawing so that eight different types of drawings have the same number of levels. Using this procedure, how many levels are assigned to each drawing type?   _____

19. The DIVIDE command in a CAD program divides a line into a specific number of equal divisions and places a symbol at each division. If the property line on a plot plan is 180 feet long, and there are 12 symbols, what is the spacing in feet between symbols?   _____

20. There are currently 256 levels active on a CAD project. If this number is reduced by half and then by half once more, how many levels will be active?   _____

21. A CAD drafter totals the number of all doors and windows in a high-rise apartment building. Within the 32 floors, there are 704 windows and 896 doors. What are the average numbers of windows and doors per floor?

                                                                            Windows _____

                                                                            Doors _____

22. A drafting department supervisor wants to determine the average number of hours that three different groups worked on their projects. Group A spent 40, 28, 32, and 36 hours on its project; group B spent 40, 40, 24, 32, and 14 hours on its project; and group C spent 40, 16, 36, 40, 28, and 32 hours on its project. Determine the average number of hours that each group worked on its project.

                                                                            Group A _____

                                                                            Group B _____

                                                                            Group C _____

CAD Problems

23. A circle contains 360 degrees. A note on a working drawing states that six holes are to be spaced at equal distances around a circle. How many degrees are in each space between the holes?   _____

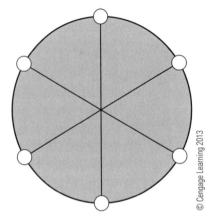

24. How many 15° sections are in this circle?   _____

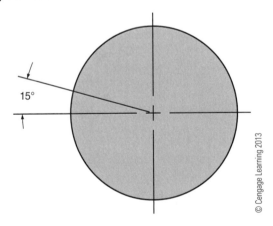

15°

25. The distances between the centers of the holes on this block are the same as the distances from each end of the block to the center of the hole nearest the end. Determine the distance between the centers.

_____

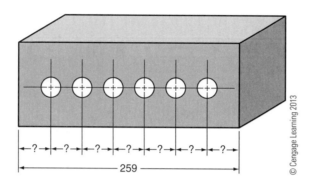

© Cengage Learning 2013

26. How many degrees are there between the locations of each hole in the figure below?

_____

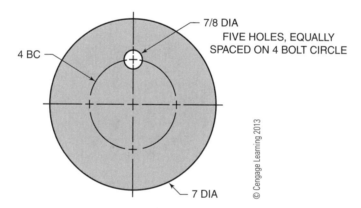

4 BC

7/8 DIA
FIVE HOLES, EQUALLY
SPACED ON 4 BOLT CIRCLE

7 DIA

© Cengage Learning 2013

27. All horizontal sections of this template are of equal width. Determine the width of each section.

Width _____

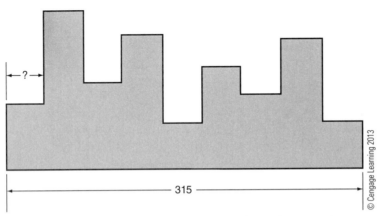

28. This CAD drawing of a house is to be scaled down by a factor of 4. What will be the length and height, in feet and inches, of the house after the scaling operation?

Length _____

Height _____

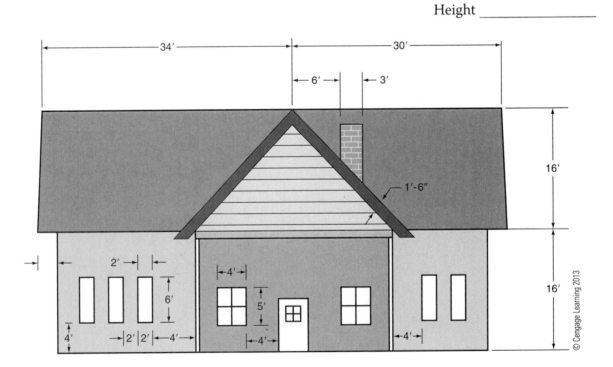

29. The CAD drawing below is to be scaled down by a factor of 4. Calculate
the overall width after the drawing of the rocker arm has been reduced. _____

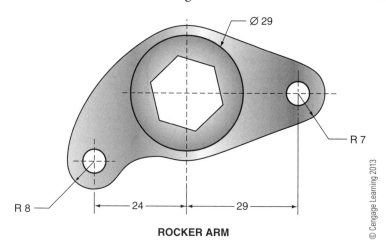

ROCKER ARM

© Cengage Learning 2013

30. This CAD drawing of an arm is to be scaled by a factor of 12. Determine
the new overall height and width of the new drawing.

Height _____

Width _____

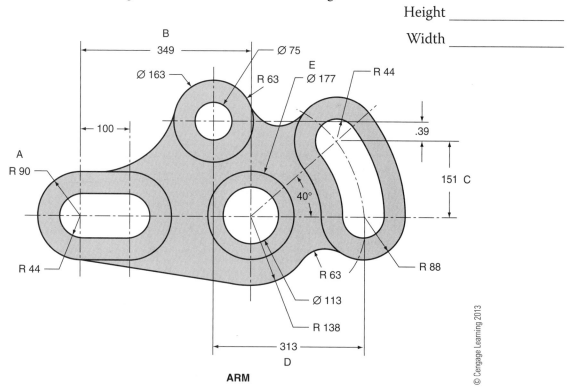

ARM

© Cengage Learning 2013

UNIT 5

Basic Principles

This unit contains a mix of problems involving the operations of addition, subtraction, multiplication, and division with whole numbers.

These basic rules (properties) apply when adding and multiplying numbers. Use these properties to make calculations easier.

	Addition	Multiplication
The Commutative Property (The Order Property)	$5 + 12 = 12 + 5$	$5 \times 12 = 12 \times 5$
The Associative Property (The Grouping Property)	$(5 + 12) + 3 = 5 + (12 + 3)$	$(5 \times 12) \times 3 = 5 \times (12 \times 3)$
The Distributive Property (Multiplication over Addition)	$5 \times (12 + 3) = (5 \times 12) + (5 \times 3)$ $(12 + 3) \times 5 = (12 \times 5) + (3 \times 5)$	

When simplifying expressions that have three or more terms, apply the following *order of operations*.

- Proceed from left to right and complete all multiplication and division steps in the order that they appear.

$$2 + 3 \times 15 - 10 = 2 + 45 - 10 \qquad \text{First multiply 3 and 15.}$$

40

- Proceed from left to right again and complete all additions and subtractions in the order that they appear.

$$2 + 45 - 10 = 47 - 10 \qquad \text{Add 2 and 45. Then subtract 10.}$$

The answer is 37.

Practical Problems

1. Drafters, working in shifts, assembled 436, 421, and 412 T-squares.

   a. How many T-squares were made in all?   **a.** _____

   b. Each shift produced 16 defective T-squares. How many products were made to specifications?   **b.** _____

2. A circle contains 12 equal sections. How many degrees are in each section?   _____

3. Sixteen CAD drafters worked for 5 days. Twenty-five architectural drafters worked for 9 days. Each work day was 8 hours long. What was the total number of hours all of the drafters worked?   _____

4. Drafters in two departments completed a package of working drawings in 1,856 hours. The nine drafters in Department X and the seven drafters in Department Y each worked an equal number of hours. How many more hours did the drafters in Department X work than the drafters in Department Y?   _____

5. An engineering department employs 268 drafters in Departments X and Z. The engineers in each department are assigned four drafters. Department X has 43 engineers. How many engineers are in division Z?   _____

6. A piece of shafting is 72″ long. Lengths of 8″, 13″, 21″, 5″, and 17″ are cut from it. Determine how much material remains after these cuts are made. (Assume that no material is lost in the cutting process.)   _____

7. A machinist can make 20 machine screws in 1 minute. Working at the same rate, how many screws can he make in 1 hour, 45 minutes, and 15 seconds?   _____

8. CAD drafters make electronic drawings on transparent overlays called *levels* or *layers*. Many CAD systems contain 256 separate levels. A specific CAD drawing uses 28 levels for construction details, 61 for floor plans, 18 for elevations, 16 for electrical symbols, 22 for furniture and appliances, and 49 for dimensions.

   a. What is the total number of levels assigned a specific function?   **a.** _____

   b. How many levels from the 256 levels that are available have not been assigned a specific function?   **b.** _____

9. A flash drive has 272,000 bytes left for file storage. Find the number of bytes remaining after the following files are saved to the drive: 36,945 bytes; 71,541 bytes; 17,612 bytes; 56,744 bytes; 27,945 bytes; and 6,815 bytes.   _____

10. In computer science, 1 byte = 8 bits. How many bits of information are contained in the following four files containing 72,396 bytes; 15,819 bytes; 56,371 bytes; and 32,250 bytes?   _____

CAD Problems

11. Square hole **A** is punched into surface **B**. The surface area of **B** is 441 square inches before hole **A** is punched. Find, in square inches, the surface area remaining after hole **A** is punched.   _____

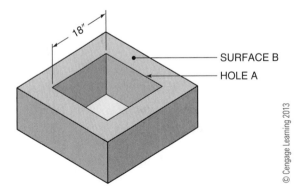

SURFACE B

HOLE A

18"

© Cengage Learning 2013

12. A *spacing collar* is used to separate adjacent parts when assembled. If the wall thickness is the same on each side, calculate the wall thickness of this spacing collar.

_____

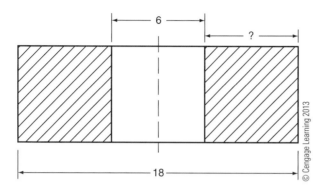

13. A drafting student centers a 4″ by 6″ block on an 8″ by 12″ sheet of paper. Determine dimensions **A** and **B**.

A _____

B _____

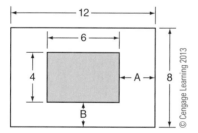

14. A shim is a tapered or wedge-shaped material. What is the perimeter of this shim?

_____

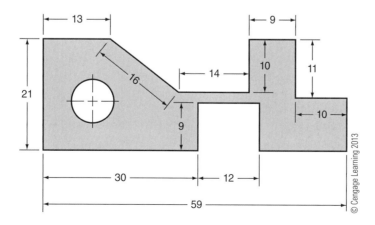

15. Calculate dimensions **A, B,** and **C** on this CAD drawing of a plate.   A _____

B _____

C _____

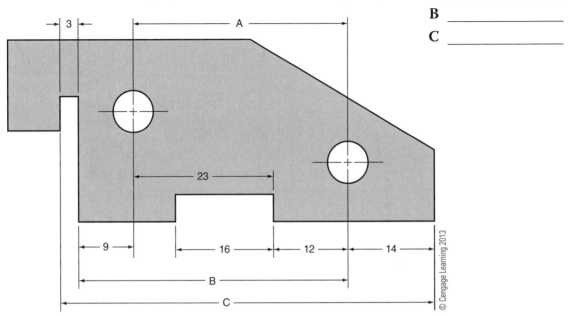

16. On this gauge, distance **AB** is equal to distance **CD**. Calculate
     distance **AB**.   _____

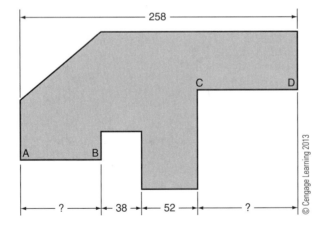

17. The sections on the drawing below, excluding the two tabs, have equal widths. What is X?

_____

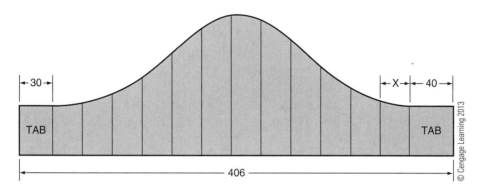

18. The 66-mm wide strip shown below is to be used to make T-guides. What is the total length in millimeters of the strip needed to make 127 T-guides?

_____

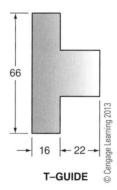

T–GUIDE

19. The CAD drawings below show two views of a wedge block. Determine lengths **A, B, C,** and **D,** in inches, from the top and front views of the block.

A ______________

B ______________

C ______________

D ______________

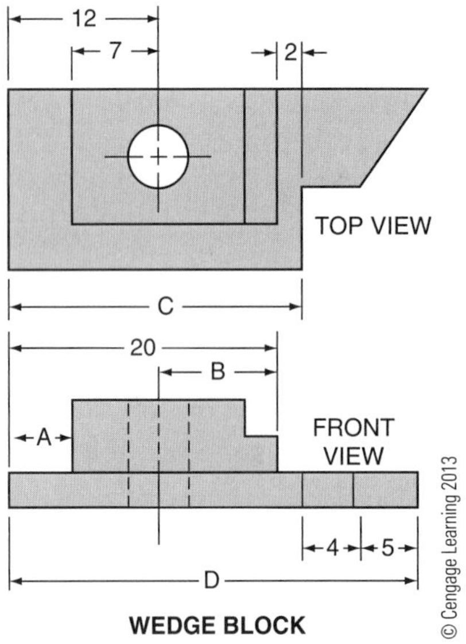
12
7
2
TOP VIEW
C
20
B
A
FRONT VIEW
4
5
D
WEDGE BLOCK
© Cengage Learning 2013

20. The CAD drawing of the house is to be scaled down by a factor of 6 along its length and by a factor of 3 along its height. Determine the new length and height in feet resulting from the scaling process.

Length _____

Height _____

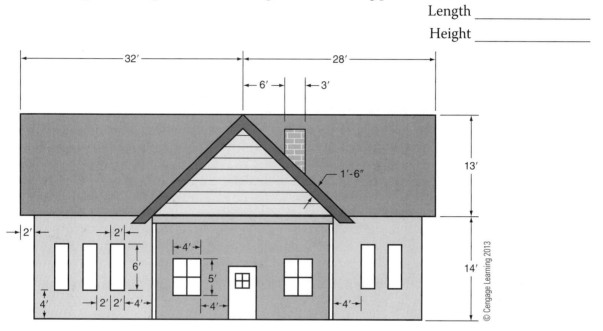

21. Determine the new overall height, width, and depth dimensions in inches if the two CAD drawings below are enlarged by a factor of 8.

Height _____

Width _____

Depth _____

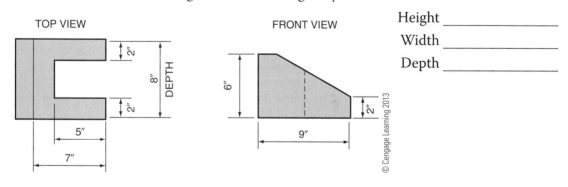

22. Use the two CAD drawings below to determine dimensions **A, B,** and **C** after they are scaled up by a factor of 6.

A _____

B _____

C _____

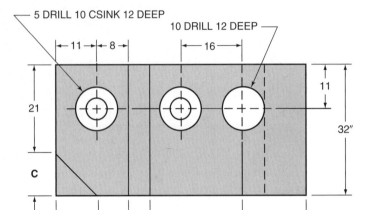

TOP VIEW

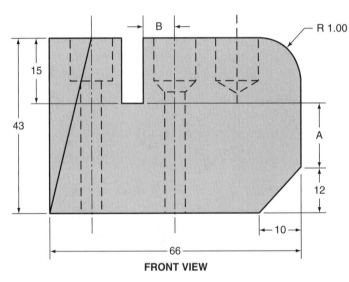

FRONT VIEW

© Cengage Learning 2013

23. The STRETCH command in a CAD program allows the CAD operator to stretch the shape of an object without affecting other crucial parts. In the CAD drawing below, the length of the object will be enlarged (stretched) by twice the distance between the center lines of holes **A** and **B.** Determine the new overall length in inches of the object after this scaling.

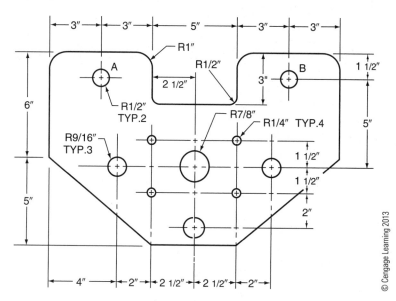

24. The CAD drawing of the plot plan below is to be scaled up by a factor of 7. Determine the new lengths in feet of *AB, BC, FG, GH, HI,* and *AI*.

AB _____

BC _____

FG _____

GH _____

HI _____

AI _____

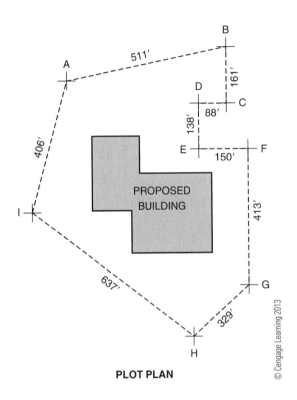

PLOT PLAN

© Cengage Learning 2013

25. Use the CAD drawings below to determine dimensions **A** through **E**.

A _____
B _____
C _____
D _____
E _____

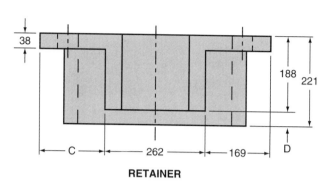

RETAINER

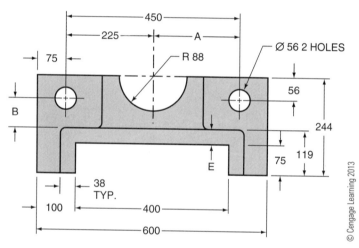

© Cengage Learning 2013

26. Use the CAD drawing shown in three dimensions below to find the new lengths for **A, B,** and **C** in millimeters, if the view is scaled down by a factor of 9.

A _____
B _____
C _____

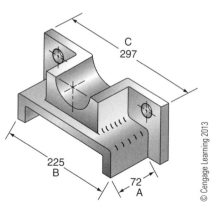

© Cengage Learning 2013

27. Use the CAD drawing of the front view of the object below and determine dimensions **A** through **E** in millimeters.

A _____

B _____

C _____

D _____

E _____

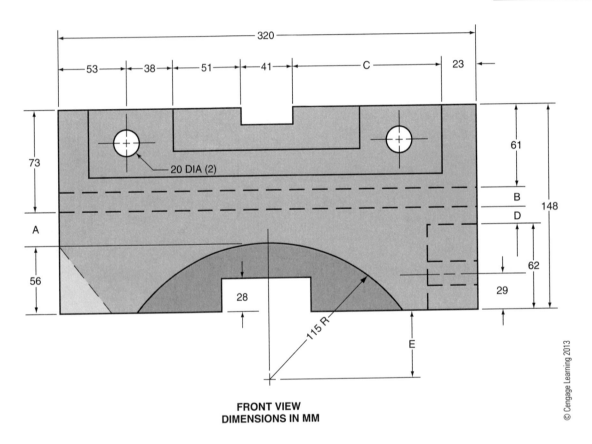

FRONT VIEW
DIMENSIONS IN MM

SECTION

Fractions

UNIT 6

Basic Principles

Accurate measurement in all drafting fields requires more precise measurements than can be accomplished using only whole numbers. This can be done by dividing a measuring unit into equal parts called fractions.

$$\frac{1}{2}, \frac{3}{4}, \frac{5}{8}, \frac{1}{16}, \frac{15}{32}, \frac{45}{64}$$

There are two parts to a fraction, the *numerator* (the number above the fraction bar) and the *denominator* (the number below the fraction bar). The denominator indicates the number of equal parts that a unit is divided into. The numerator indicates the number of equal parts.

There are two types of fractions, proper and improper.
- *Proper fractions* are less than 1; their numerators are less than their denominators. Examples: ¾, ⅞, ⁵⁄₁₆
- *Improper fractions* are greater than or equal to 1; their numerators are greater than or equal to their denominators. Examples ⁴⁄₄, ⁹⁄₈, ²⁵⁄₁₆

To add two or more fractions, the denominators must be the same. Then you can add the numerators.

NOTE: The denominator of the sum is the same as the denominators of the like fractions being added.

EXAMPLE: $\dfrac{3}{16} + \dfrac{5}{16} + \dfrac{7}{16}$

You can write the addends one under the other like this.

$$\begin{array}{r} \dfrac{3}{16} \\[6pt] \dfrac{5}{16} \\[6pt] +\dfrac{7}{16} \\[6pt] \hline \dfrac{15}{16} \end{array}$$

The sum is $^{15}\!/_{16}$.

If the denominators of the fractions to be added are not the same, you need to find a common denominator.

EXAMPLE: $\dfrac{1}{2} + \dfrac{1}{3} + \dfrac{1}{8}$

METHOD 1

One way to find a common denominator is to multiply all the denominators.

STEP 1:   Multiply 2, 3, and 8.

$2 \times 3 \times 8 = 48$.

So, 48 is a common denominator.

STEP 2:   Rewrite the addends as equivalent fractions with denominators of 48. Add the numerators.

$$\begin{array}{r} \dfrac{1}{2} = \dfrac{1}{2} \times \dfrac{24}{24} = \dfrac{24}{48} \\[10pt] \dfrac{1}{3} = \dfrac{1}{3} \times \dfrac{16}{16} = \dfrac{16}{48} \\[10pt] +\dfrac{1}{8} = \dfrac{1}{8} \times \dfrac{6}{6} = +\dfrac{6}{48} \\[6pt] \hline \dfrac{46}{48} \end{array}$$

STEP 3:   Simplify the sum by dividing the numerator and denominator by their GCF, 2.

$$\frac{46 \div 2}{48 \div 2} = \frac{23}{24}$$

The sum is $^{23}/_{24}$.

METHOD 2

Another way to add fractions with unlike denominators is to find the *least* common denominator (LCD). The LCD is the smallest number that all the denominators divide into evenly.

EXAMPLE: $\dfrac{1}{2} + \dfrac{1}{3} + \dfrac{1}{8}$

STEP 1:   To find the LCD for 2, 3, and 8, list the multiples of each denominator. Then circle the least multiple that is common to each denominator (LCM). The LCM in this example is 24, so the LCD is 24.

Denominators	Multiples
2	2, 4, 6, 8, 10, 12..., 22, ㉔ 26, 28...
3	3, 6, 9, 12, 15, 18, 21, ㉔ 27, 30...
8	8, 16, ㉔ 32, 40...

STEP 2:   Multiply the numerator and denominator of each fraction by 1 written as an improper fraction so that the denominator of the equivalent fractions will be 24. In this example, the improper fractions that equal 1 are $^{12}/_{12}$, $^{8}/_{8}$, and $^{3}/_{3}$.

STEP 3: Multiply to write the equivalent fraction for each addend. Then, add the numerators and keep the same denominator in the sum.

$$\frac{1}{2} = \frac{1}{2} \times \frac{12}{12} = \frac{12}{24}$$

$$\frac{1}{3} = \frac{1}{3} \times \frac{8}{8} = \frac{8}{24}$$

$$+\frac{1}{8} = \frac{1}{8} \times \frac{3}{3} = +\frac{3}{24}$$

$$\frac{23}{24}$$

The sum is $^{23}/_{24}$.

The addends in the following example include the mixed numbers 2⅓ and 1⅕. A mixed number is the sum of a whole number and a fraction: 2⅓ = 2 + ⅓ and 1⅕ = 1 + ⅕.

RULE: To add fractions and mixed numbers,

• add the fraction parts
• add the whole number parts
• write the sum in simplest terms as necessary.

EXAMPLE: $2\frac{1}{3} + 1\frac{1}{5} + \frac{1}{6}$

STEP 1: The LCD of the fractions is 30, the least common multiple (LCM) of 3, 5, and 6.

STEP 2: Rewrite the fraction in each mixed number as an equivalent fraction with a denominator of 30.

STEP 3:   Add the fractions and the whole numbers.

$$2\frac{1}{3} \times \frac{10}{10} = 2\frac{10}{30}$$

$$1\frac{1}{5} \times \frac{6}{6} = 1\frac{6}{30}$$

$$+2\frac{1}{6} \times \frac{5}{5} = +2\frac{5}{30}$$

$$5\frac{21}{30}$$

The sum is $5^{21}/_{30}$, which can be simplified to $5^{7}/_{10}$.

NOTE: In many calculators, you can use the [a%] key to enter a fraction or a mixed number.

- If a fraction is a proper or improper fraction, press the [a%] key after entering the numerator of each fraction. The sum is shown as a fraction.

EXAMPLE: $\dfrac{3}{16} + \dfrac{1}{2} = \dfrac{11}{16}$

    3 [a%] 16 [+] 1 [a%] 2 [=]

- If a number is a mixed number, press the [a%] key after entering the whole number and after entering the numerator of each fraction. The sum is shown as a mixed number.

EXAMPLE: $5\dfrac{11}{32} + 3\dfrac{15}{64} = 8\dfrac{37}{64}$

    5 [a%] 11 [a%] 32 [+] 3 [a%] 15 [a%] 64 [=]

Converting Between Feet and Inches

When working in related drafting fields, you will encounter measurements given in feet and inches or just inches alone. These may need to be converted.
- To convert feet and inches to inches, multiply the number of feet by 12 in./ft. and then add the number of inches to the product.
- To convert inches to feet and inches, divide the number of inches by 12 in./ft. to find the number of feet. The remainder equals the number of inches.

EXAMPLE: Express 3 feet 7¾ inches in inches.

NOTE: In drafting, a measurement such as 3 feet 7¾ inches is written as 3'- 7¾". A hyphen separates the different units.

STEP 1:   Convert 3 feet to inches. (The units in ft. cancel out.)

3 ft. × 12 in./ft. = 36 in.

STEP 2:   Add the inches.

36 in. + 7¾ in. = 43¾ in.

EXAMPLE: Express 145½ inches in feet and inches.

Divide the number by 12 in./ft. and add the remainder of 1 in. to ½ in.

$$12 \overline{)145\tfrac{1}{2}} \quad \longleftarrow \text{ Quotient in feet} = 12$$

$$\begin{array}{r} 12 \\ 12\,\overline{)145\tfrac{1}{2}} \\ -12 \\ \hline 25 \\ -24 \\ \hline 1\tfrac{1}{2} \end{array} \quad \longleftarrow \text{ Remainder in inches}$$

The answer is 12 ft. 1½ in. This can also be written as 12'-1½"

Skill Problems

Calculate the sums. Express all answers in simplest form and include the unit for each sum.

1. ⅜ inch + ⅝ inch   _____

2. ¼ inch + ½ inch   _____

3. ⁵⁄₁₆ inch + ⁹⁄₁₆ inch   _____

4. ⅛ inch + ³⁄₁₆ inch   _____

5. 2¼ inches + 3³⁄₁₆ inches + 4⅝ inches   _____

6. 3½ hours + 2¼ hours + 1¾ hours   _____

7. 4⅓ hours + ½ hour + 3⅝ hours   _____

8. 2⁷⁄₁₆ inches + 1 inch + 3²¹⁄₃₂ inches   _____

9. 6⅔ yards + 4½ yards + 1¼ yards   _____

10. 5¹⁵⁄₁₆ inches + 3³⁄₃₂ inches + 1⁴⁷⁄₆₄ inches   _____

Practical Problems

11. Determine the total length in inches of a line formed by measurements
    of 2⅛", 1¾", 3½", and 1⅜"   _____

12. Calculate, in inches, the overall length of a shaft made up of five
    sections whose lengths are 3⁵⁄₁₆", 2⅞", 1¾", 1⁹⁄₃₂", and 6½".   _____

13. What is the sum of 4⅜", 3¹⁄₁₆", 11¹⁄₁₆", and 3½"?   _____

14. What length of stock, in inches, is needed to make a shaft with linear
    dimensions of 1⁵⁄₃₂", 1³⁄₁₆", 2⁷⁄₃₂", 3⁹⁄₁₆", ⅝", and ¾"?   _____

CAD Problems

15. Calculate the length of this flat-head cap screw.   _____

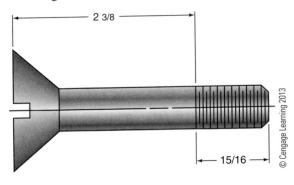

2 3/8

15/16

© Cengage Learning 2013

16. What is the total length of this link?

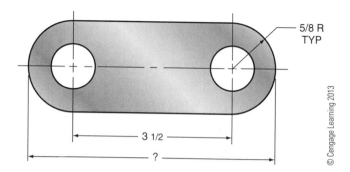

5/8 R
TYP

3 1/2

?

© Cengage Learning 2013

17. What are the lengths of dimension **A** and dimension **B** on this template?

A _____

B _____

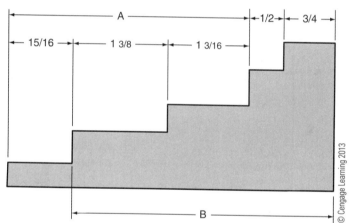

A

1/2 3/4

15/16   1 3/8   1 3/16

B

© Cengage Learning 2013

18. Calculate the total length of this shaft.

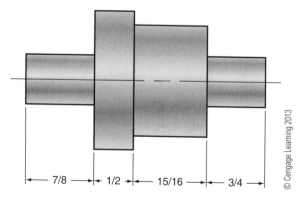

7/8   1/2   15/16   3/4

© Cengage Learning 2013

19. Determine dimensions **A** and **B** on this strap.

A _____

B _____

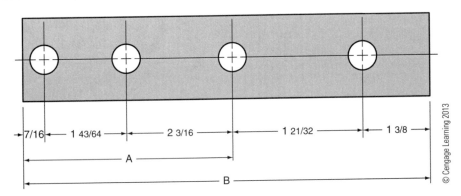

20. Calculate dimensions **A** and **B** on this locator block.

A _____

B _____

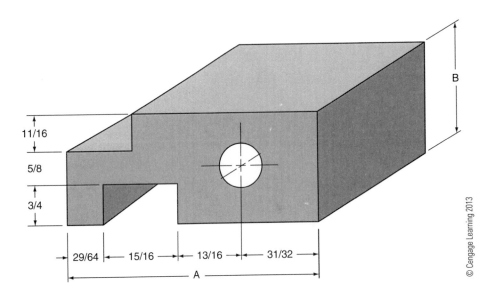

21. Use the diagram below to determine dimension **F** if **A** has a measure of 1¹⁄₁₆, **G** has a measure of 3⅝, and **E** has a measure of 2²⁷⁄₃₂.

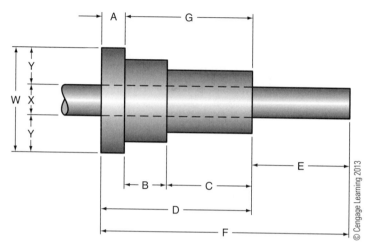

© Cengage Learning 2013

22. Property line lengths are usually expressed in feet and inches. Determine in feet and inches the perimeter of the plot plan below using the given distances.

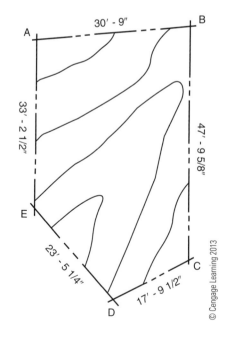

© Cengage Learning 2013

23. What is the overall length of the arm in this CAD drawing? _____

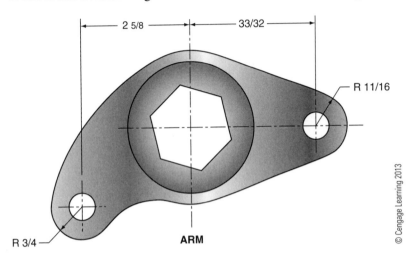

2 5/8   33/32

R 11/16

R 3/4   **ARM**

© Cengage Learning 2013

24. Determine dimensions **A** and **B** in the CAD drawing of the link below.   A _____

B _____

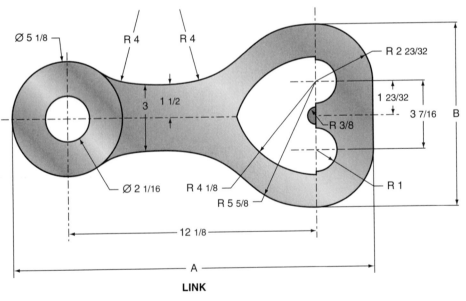
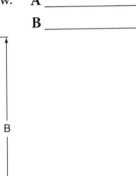

Ø 5 1/8   R 4   R 4   R 2 23/32

1 1/2

3   1 23/32   3 7/16   B

R 3/8

Ø 2 1/16   R 4 1/8   R 1

R 5 5/8

12 1/8

A

LINK

© Cengage Learning 2013

25. Use this CAD drawing to calculate the height and width of this figure
    in inches.

Height _____

Width _____

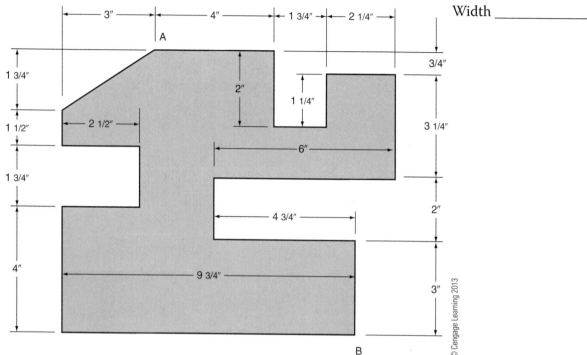

26. Use the CAD drawing above to calculate the sum of the line lengths
    from point **A** to point **B** in inches, moving in a clockwise direction.

    _____

27. Use this CAD drawing of a cam to determine dimension A.

    _____

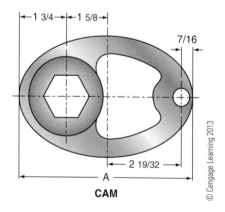

CAM

28. Determine dimension **F** in inches if **A** has a measure of ⁷⁄₁₆", **B** has a measure of ⁴⁷⁄₆₄", **C** has a measure of 1¾", and **E** has a measure of 2⅜". _____

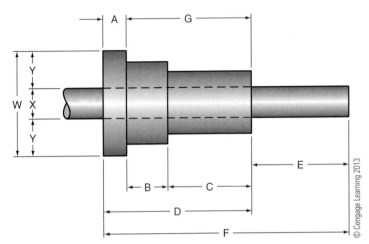

© Cengage Learning 2013

29. Use the CAD drawing below to calculate dimensions **A** and **B** and the total height and width of the figure.

A _____

B _____

Height _____

Width _____

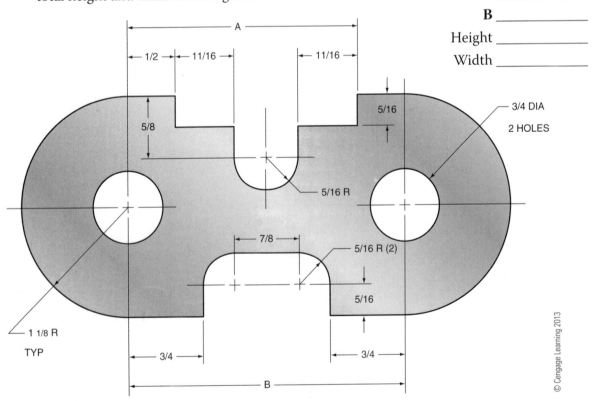

© Cengage Learning 2013

UNIT 7

Basic Principles

As in the addition of fractions, to subtract two fractions, their denominators must be the same.

RULE:

- If the denominators are the same, subtract the numerators. The denominator of the difference is the same as the denominators of the fractions you subtracted.
- If the denominators are not the same, find a common denominator or the least common denominator (LCD).
- Rewrite the fraction(s) in the problem as equivalent fractions with the common denominator. Subtract the numerators. The denominator of the difference is the common denominator.
- In either case, simplify the difference as necessary.

EXAMPLE: $\dfrac{7}{8} - \dfrac{19}{32}$

STEP 1: Multiply ⅞ by 4/4 to write it as an equivalent fraction with a denominator of 32.

$$\frac{7}{8} \times \frac{4}{4} = \frac{28}{32}$$

67

STEP 2:   Subtract the numerators. Keep the denominator the same.

$$\frac{28}{32} - \frac{19}{32} = \frac{9}{32}$$

The difference is $\frac{9}{32}$.

7 a% 8 − 19 a% 32 =

EXAMPLE: $\dfrac{15}{16} - \dfrac{3}{4}$

STEP 1:   Write ¾ as an equivalent fraction with a denominator of 16.

$$\frac{3}{4} \times \frac{4}{4} = \frac{12}{16}$$

STEP 2:   Subtract the numerators.

$$\frac{15}{16} - \frac{12}{16} = \frac{3}{16}$$

The difference is $\frac{3}{16}$.

15 a% 16 − 3 a% 4 =

Skill Problems

Subtract as indicated. Express all answers in simplest form.

1. $\frac{3}{4}$ in.
    $-\frac{3}{8}$ in.

2. $10\frac{3}{8}$ lb.
    $-3\frac{1}{4}$ lb.

3. $1\frac{7}{8}$ yd.
    $-\frac{5}{16}$ yd.

4. 21 ft.
    $-\frac{2}{3}$ ft.

5. $^{15}\!/_{16}$ inch $- {}^{15}\!/_{24}$ inch

6. $4\frac{3}{8}$ lb. $- 1\frac{1}{4}$ lb.

7. $3\frac{1}{4}$ lb. $- 1\frac{1}{2}$ lb.

8. $2^{13}\!/_{64}$ inches from $3\frac{5}{8}$ inches

9. $\frac{1}{8}$ foot from $1\frac{3}{4}$ foot

10. $1\frac{1}{2}$ pounds from $4\frac{3}{8}$ pounds

Practical Problems

11. The thickness of a steel block is $^{27}\!/_{64}$ inches. A milling machine is used to remove a $^{3}\!/_{16}$-inch cut from the block. How thick, in inches, is the remaining block?

12. A $^{3}\!/_{32}$-inch cut is removed from a $^{31}\!/_{64}$-inch block. What is the resulting thickness of the block in inches?

13. A CAD drafter spends 80 hours on three projects. If she spends $61\frac{3}{4}$ hours working at the same rate on the projects, how many more hours will it take her to finish the projects?

14. The requirement for the sum of three measurements on a drawing should be $7\frac{7}{8}$″. The individual measurements shown on the drawing are $1\frac{1}{2}$″, $3\frac{1}{8}$″, and $2^{11}\!/_{16}$″. What is the difference between the sum of the actual measurements and the sum of the required measurements?

15. What is the difference in diameters between a $5\frac{1}{4}$-foot wheel and a $3\frac{1}{2}$-foot wheel?

CAD Problems

16. Calculate the thickness of the wall around this collar.   _____

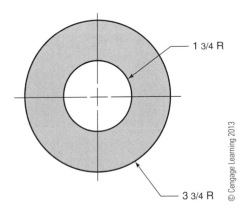

1 3/4 R

© Cengage Learning 2013

3 3/4 R

17. Each hole on this plate has a radius of ⁹⁄₁₆. Determine dimension **A**.   _____

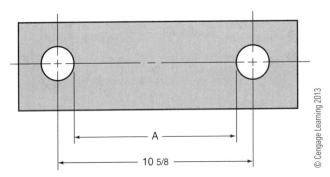

A

10 5/8

© Cengage Learning 2013

18. Determine the inside diameter of this washer.   _____

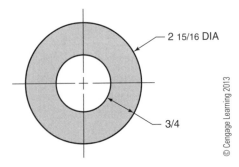

2 15/16 DIA

3/4

© Cengage Learning 2013

19. Determine dimensions **C** and **D** on this shim.

C _____

D _____

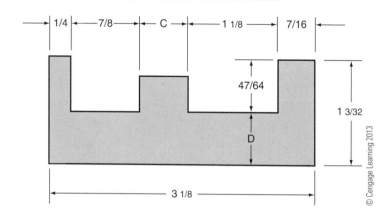

20. a. Determine dimension **A** on this template.

b. Calculate the difference between the vertical dimensions.

a. _____

b. _____

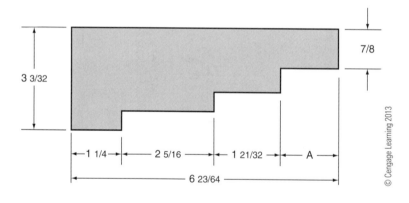

21. Calculate dimensions **A** and **B** on this strap.

A _____

B _____

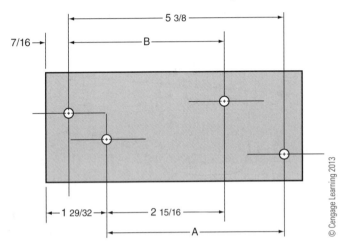

© Cengage Learning 2013

22. a. What is the depth of the drilled hole in this diagram?

a. _____

b. If a hole is first drilled to a depth of $1^{47}\!/_{64}$, how much deeper must it
be drilled to reach the bottom?

b. _____

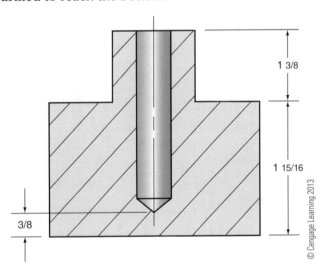

© Cengage Learning 2013

23. In the diagram below, dimension **A** is ⅝″, dimension **B** is 1¹⁄₁₆″, dimension **C** is 3¹³⁄₃₂″, and dimension **F** is 9⁴¹⁄₆₄″. What is dimension **E**? _____

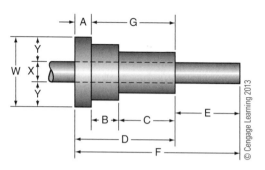

© Cengage Learning 2013

24. Use the CAD drawing of this floor plan and determine dimensions **A, B,** and **C.**

A _____

B _____

C _____

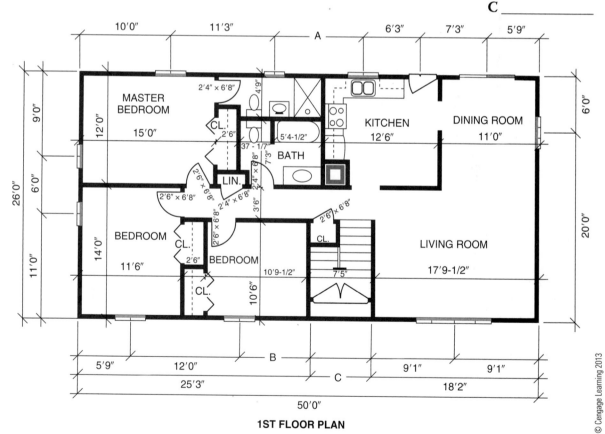

1ST FLOOR PLAN

© Cengage Learning 2013

25. Determine dimensions **A** and **B** on the CAD drawing below.

A _____

B _____

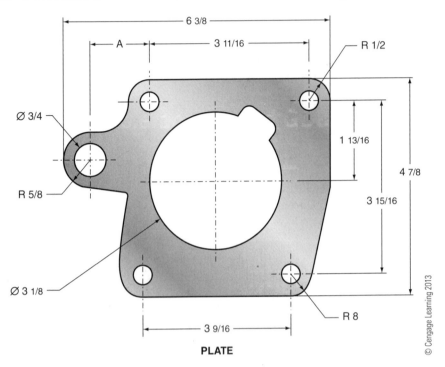

PLATE

© Cengage Learning 2013

26. Calculate dimensions **A** and **B** on this CAD drawing

A _____

B _____

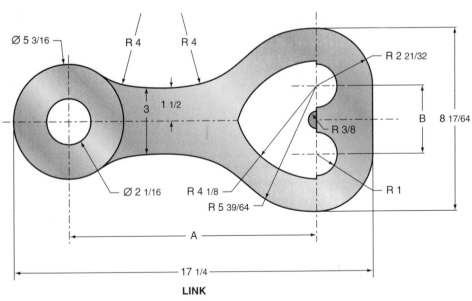

LINK

© Cengage Learning 2013

27. Use this CAD drawing of a gasket and calculate dimensions **A** and **B**.

A _____

B _____

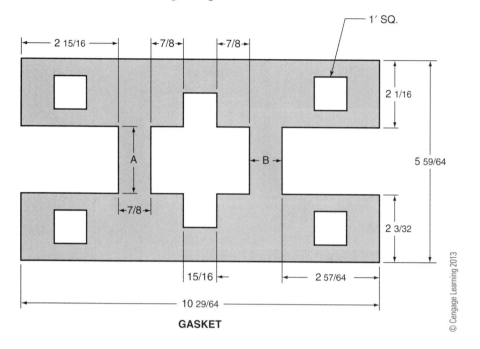

GASKET

28. This CAD drawing shows a section of wall. What is dimension **A** in feet and inches if the height from the bottom of the footing to the top of the subfloor is 2'-8¾"?

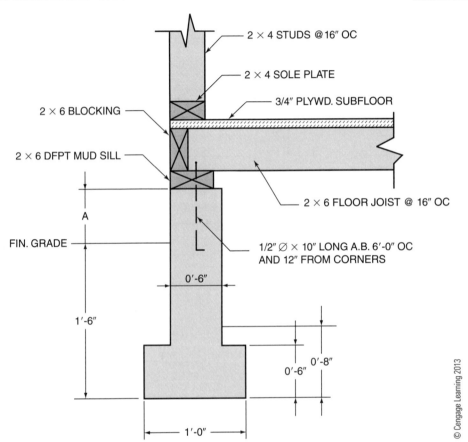

2 × 4 STUDS @ 16" OC

2 × 4 SOLE PLATE

2 × 6 BLOCKING

3/4" PLYWD. SUBFLOOR

2 × 6 DFPT MUD SILL

A

FIN. GRADE

2 × 6 FLOOR JOIST @ 16" OC

1/2" Ø × 10" LONG A.B. 6'-0" OC
AND 12" FROM CORNERS

0'-6"

1'-6"

0'-8"

0'-6"

1'-0"

29. Calculate dimensions **A, B,** and **C** on the two CAD drawings of a bracket.

A _____

B _____

C _____

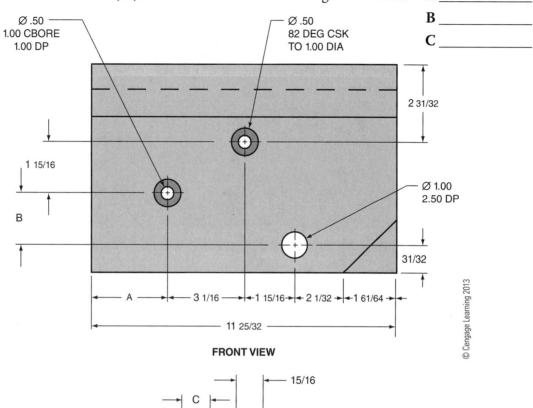

FRONT VIEW

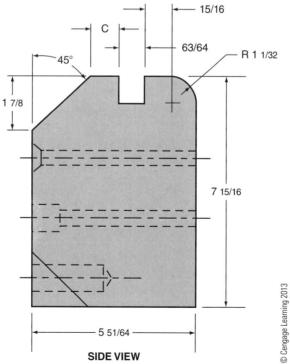

SIDE VIEW

30. Use this CAD drawing and determine dimensions **A** and **B**.

A _____

B _____

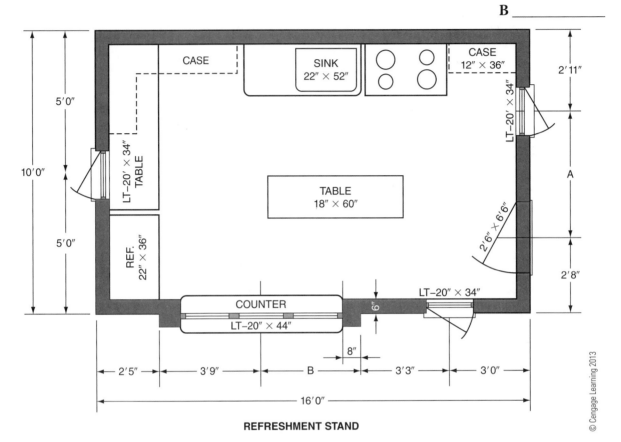

REFRESHMENT STAND

31. Use this CAD drawing and calculate the five missing dimensions
    A through **E**.

A _____

B _____

C _____

D _____

E _____

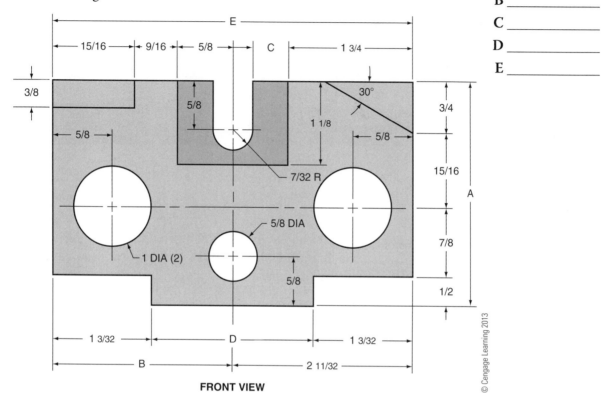

FRONT VIEW

© Cengage Learning 2013

UNIT 8

Multiplication

Basic Principles

To multiply fractions, multiply their numerators and their denominators. Then write the product of the numerators over the product of the denominators. Simplify the fraction if necessary by reducing it to simplest form. If the resulting product is an improper fraction, rewrite it as a mixed number in simplest form

EXAMPLE: $\dfrac{3}{8} \times \dfrac{4}{5}$

STEP 1: Multiply the numerators and denominators.

$$\frac{3}{8} \times \frac{4}{5} = \frac{3 \times 4}{8 \times 5} = \frac{12}{40}$$

STEP 2: Simplify the product by dividing the numerator and denominator by their greatest common factor (GCF), which is 4.

$$\frac{12 \div 4}{40 \div 4} = \frac{3}{10}$$

The product is 3/10.

NOTE: You can simplify a multiplication problem involving fractions in two ways.

- Multiply the factors and then simplify the product by dividing its numerator and denominator by their greatest common factor (GCF), as seen in the example above,

80

- Or before multiplying, divide a numerator and a denominator in two factors by their GCF. In the example above, 8 and 4 have a GCF of 4, so before multiplying, the factors can be simplified to ½ and ⅕. The product is the same.

$$\frac{3}{_2 8} \times \frac{4^1}{5} = \frac{3}{10}$$

RULE: To multiply a fraction and a whole number, write the whole number as an improper fraction whose denominator is 1. Then multiply the numerators and denominators. If the product is an improper fraction, divide the numerator by the denominator and write the fraction as a mixed number.

EXAMPLE: $\dfrac{3}{16} \times 7$

$$\frac{3}{16} \times 7 = \frac{3}{16} \times \frac{7}{1} = \frac{3 \times 7}{16 \times 1} = \frac{21}{16} = 1\frac{5}{16}$$

3 [a%] 16 [×] 7 [=]

RULE: To multiply a fraction and a mixed number, rewrite the mixed number as an improper fraction first. Then, multiply the numerators and denominators. If the product is an improper fraction, write the product as a mixed number.

EXAMPLE: $\dfrac{3}{4} \times 2\dfrac{1}{2}$

$$\frac{3}{4} \times 2\frac{1}{2} = \frac{3}{4} \times \frac{5}{2} = \frac{3 \times 5}{4 \times 2} = \frac{15}{8} = 1\frac{7}{8}$$

3 [a%] 4 [×] 5 [a%] 2 [=]

Skill Problems

Multiply. Express all products in simplest form as necessary.

1. ⅜ × ½   _____

2. ³⁄₁₆ × ¼   _____

3. ¼ × ⁷⁄₁₆   _____

4. 9 × ⁵⁄₁₆   _____

5. 1⅝ × ⅓   _____

6. ⅔ × ³⁄₇   _____

7. 2½ × 7   _____

8. 3½ × 1¼ × 1⅜   _____

9. 1½ × ¾ × 3⅛   _____

Practical Problems

10. Seven pieces of copper are to be sheared from one length. Each piece
    is 2¾ inches long. What is the length, in inches, of the copper in each
    piece if there is no waste?   _____

11. Cap screws are made from ⅜″ diameter bar stock. Each cap screw is
    1¾″ long. What length bar, in inches, is needed to make 24 cap screws? _____

12. The thickness of a washer is ³⁄₃₂″. What is the height in inches of a stack
    of 24 washers?   _____

13. Drafting vellum costs 7½ cents per sheet. How much do 90 sheets
    cost?

14. The area of a rectangle can be found by multiplying its length and its width. What is the area of this rectangle in square inches?   _____

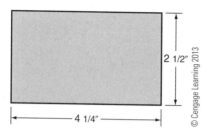

15. A box of 12-inch drafting scales weighs 3¾ pounds. A dealer has 20½ boxes in stock. What is the total weight in pounds of the scales?   _____

16. Isometric projections and isometric drawings are different ways to represent three-dimensional objects. An isometric projection is about ¾ the size of an isometric drawing. A 6-inch line is drawn on an isometric drawing. About how many inches long is the corresponding line on the isometric projection?   _____

17. A CAD drafter lays off 12 line segments of ¹³⁄₁₄ inches each. Calculate the total length, in inches, of these line segments.   _____

18. A CAD drafter spends a total of 9¼ work days on an architectural design project. During five work days, he spent 9½ hours each day working on the project. For the remaining work days, he spent 8½ hours on the project. What was the total number of hours that the drafter spent working on this project?   _____

19. If ¼ inch on an architectural drawing represents 1 foot, how many inches on the drawing represent 45 feet?   _____

20. Calculate the total cost of a pack of 50 CD-ROM discs if one CD costs 97½ cents.   _____

CAD Problems

21. Determine the overall length of this sheet metal development.   _____

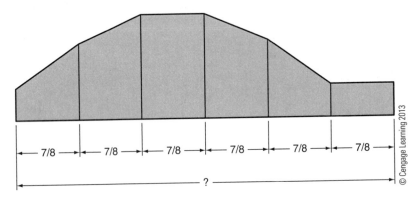

22. All holes on this plate are equally spaced. Determine the distance
     between the center of hole **A** and the center of hole **B**.   _____

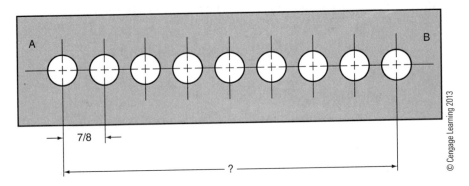

23. Calculate the distance between floors and the length of dimension **A.**

Distance between floors _____

Dimension A _____

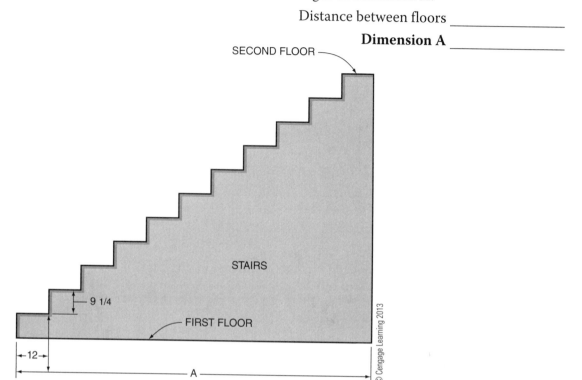

SECOND FLOOR

STAIRS

9 1/4

FIRST FLOOR

12

A

© Cengage Learning 2013

24. What would be the lengths of the radii in circles **A**, **B**, and **C** on the
CAD drawing below if the gasket is enlarged by a factor of 5?

A _____

B _____

C _____

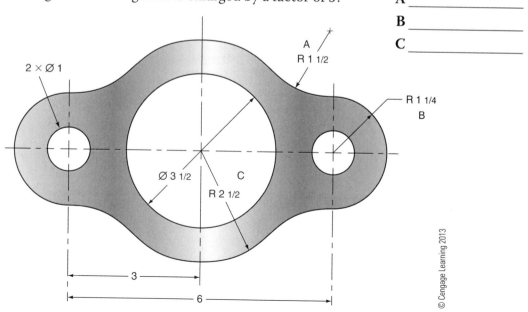

2 × Ø 1

A
R 1 1/2

R 1 1/4
B

Ø 3 1/2   C

R 2 1/2

3

6

© Cengage Learning 2013

25. The CAD drawing below will be inserted into another project as a symbol. Determine the new dimensions of **A**, **B**, and **C** if the drawing is increased by a factor of 7 when inserted.

A _____

B _____

C _____

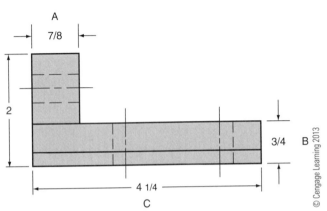

26. Use the CAD drawing below and calculate dimensions **A** and **B** if the drawing is scaled up by a factor of 3.

A _____

B _____

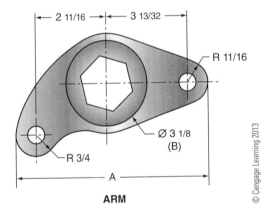

ARM

27. Calculate the new overall dimensions for the parts illustrated in the CAD drawings below if they are to be scaled up by a factor of 8.

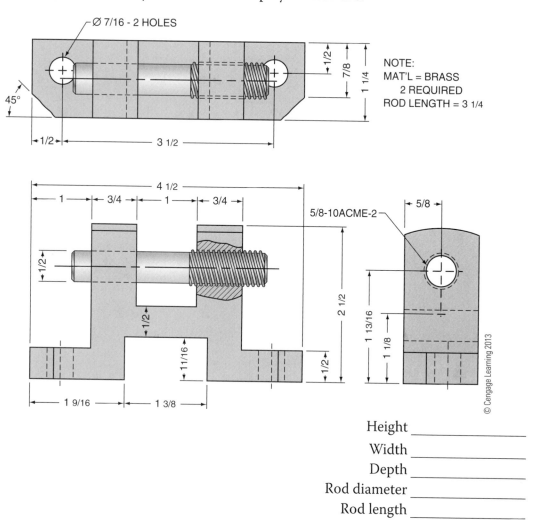

Height _____

Width _____

Depth _____

Rod diameter _____

Rod length _____

28. The CAD drawing of the automotive gasket below needs to be enlarged
to 5 times its size. Determine the dimensions of features **A** through **D**
when the gasket is enlarged.

A _____

B _____

C _____

D _____

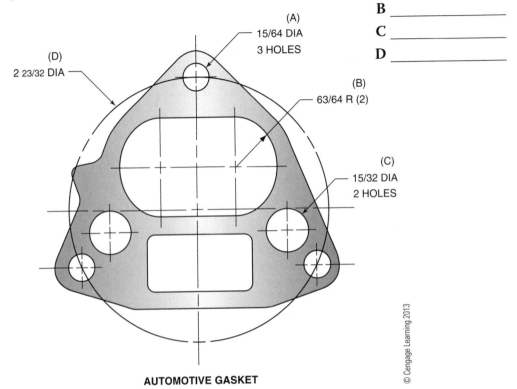

(A)
15/64 DIA
3 HOLES

(D)
2 23/32 DIA

(B)
63/64 R (2)

(C)
15/32 DIA
2 HOLES

AUTOMOTIVE GASKET

UNIT 9

Division

Basic Principles

Division is the inverse (opposite) operation of multiplication. But to divide fractions, you need to identify the *reciprocal* of the divisor and then rewrite the problem as a multiplication problem. The reciprocal of any number, except 0, is the number turned "upside down." All numbers other than 0 have reciprocals. The product of a number and its reciprocal is 1. Study these examples.

Number	Reciprocal	Product
$\dfrac{1}{2}$	$\dfrac{2}{1}$	$\dfrac{1}{2} \times \dfrac{2}{1} = 1$
$\dfrac{32}{8}$	$\dfrac{8}{32}$	$\dfrac{32}{8} \times \dfrac{8}{32} = 1$
$\dfrac{5}{16}$	$\dfrac{16}{5}$	$\dfrac{5}{16} \times \dfrac{16}{5} = 1$
3	$\dfrac{1}{3}$	$3 \times \dfrac{1}{3} = 1$

EXAMPLE: $\dfrac{3}{4} \div \dfrac{5}{16}$

STEP 1: Rewrite the divisor as the reciprocal of ⁵⁄₁₆. The reciprocal of ⁵⁄₁₆ is ¹⁶⁄₅.

STEP 2: Write the division problem as a multiplication problem using the reciprocal as the second factor.

NOTE: The first factor, ¾, stays the same.

$$\dfrac{3}{4} \div \dfrac{5}{16} = \dfrac{3}{4} \times \dfrac{16}{5}$$

STEP 3: Reduce like factors in the numerator ¾ and in the numerator ¹⁶⁄₅. Then multiply to calculate the product.

$$\dfrac{3}{{}_1 4} \times \dfrac{16^4}{5} = \dfrac{12}{5}$$

STEP 4: Write the improper fraction as a mixed number.

$$\dfrac{12}{5} = 2\dfrac{2}{5}$$

So, ¾ (divided by symbol) ⁵⁄₁₆ $= 2\dfrac{2}{5}$.

NOTE: If using a calculator, it is not necessary to enter the reciprocal of the divisor and then multiply. Just use the division key.

3 [a%] 4 [÷] 5 [a%] 16 [=]

EXAMPLE: $\dfrac{1}{4} \div \dfrac{3}{8}$

$$\dfrac{1}{4} \div \dfrac{3}{8} = \dfrac{1}{4} \times \dfrac{8}{3}$$

$$= \dfrac{8}{12}$$

$$= \dfrac{2}{3}$$

1 [a%] 4 [÷] 3 [a%] 8 [=]

EXAMPLE: $3 \div \dfrac{1}{5}$

$$3 \div \frac{1}{5} = \frac{3}{1} \times \frac{5}{1}$$

$$= \frac{15}{1}$$

$$= 15$$

 $3 \div 1$ a% $5 =$

EXAMPLE: $2\dfrac{1}{8} \div 1\dfrac{1}{2}$

Rewrite each mixed number as an improper fraction.

$$2\frac{1}{8} = \frac{17}{8} \qquad \text{and} \qquad 1\frac{1}{2} = \frac{3}{2}$$

Rewrite the division problem. Then, multiply and simplify.

$$2\frac{1}{8} \div 1\frac{1}{2} = \frac{17}{8} \div \frac{3}{2}$$

$$= \frac{17}{\underset{4}{8}} \times \frac{\overset{1}{2}}{3}$$

$$= \frac{17}{12}$$

$$= 1\frac{5}{12}$$

 2 a% 1 a% $8 \div 1$ a% 1 a% $2 =$

Skill Problems

Divide.

NOTE: When dividing like units, the units cancel out and the answer has no unit.

1. ⅞ inch ÷ ⁷⁄₃₂ inch   _____

2. 5½ lb. ÷ 3¾ lb.   _____

3. 17 yd. ÷ ⅝ yd.   _____

4. $^{15}\!/_{16}$ inch ÷ 4   _____

5. $^{13}\!/_{16}$ inch ÷ $^1\!/_8$ _____

6. $5^3\!/_4$ yd. ÷ $^2\!/_3$ _____

7. $^3\!/_7$ ÷ $4^2\!/_3$ _____

8. $2^3\!/_{16}$ inch ÷ $1^1\!/_2$ inch   _____

9. $2^7\!/_{10}$ cm ÷ $6^3\!/_{10}$ cm   _____

10. $3^1\!/_4$ ÷ $^3\!/_5$ _____

Practical Problems

11. A regular octagon has eight equal sides. If the perimeter of a regular
    octagon is $11^3\!/_8$ inches, what is the length of each side in inches?   _____

12. A piece of drafting lead is $37^5\!/_8$ inches long. How many $5^3\!/_8$-inch drafting
    leads can be cut from it? (Assume no allowance is made for waste.)   _____

13. How many pieces of metal, each $^{27}\!/_{64}$″ long, can be made from a strip
    54″ long? Assume there is no waste in cutting.   _____

14. A CAD drafter completes a certain job in $^2\!/_3$ hour. If he works at the
    same rate, how many jobs of the same kind can he complete in 4 hours?   _____

15. A stack of $^5\!/_{16}$″ thick erasers is $4\,^{11}\!/_{16}$″ high. How many erasers are in the
    stack?   _____

16. A regular hexagon has six equal sides. The perimeter of a regular
    hexagon is $19^1\!/_2$″. Determine, in inches, the length of each side of the
    hexagon.   _____

17. On a drawing, $^1\!/_8$ inch equals 1 inch.

    a. How many inches are represented by a line that is $3^3\!/_{16}$ inches long?   **a.** _____

    b. How many inches are represented by a line that is $3^{13}\!/_{32}$ inches long?   **b.** _____

18. A line that is $18^3\!/_4$″ long is to be divided into 15 equal segments. How
    long is each new segment in inches?   _____

CAD Problems

19. All steps on this template are equal. Calculate dimension **A**.

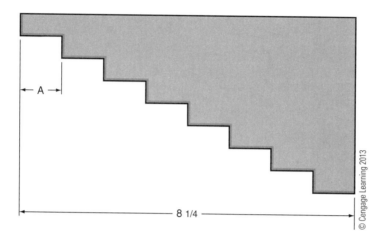

© Cengage Learning 2013

8 1/4

20. All nine segments of this sheet metal layout of a truncated cylinder are equal in width. Determine dimension **A**.

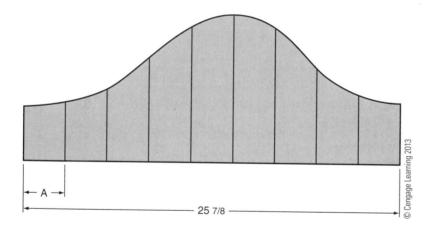

© Cengage Learning 2013

25 7/8

21. A brass rod that is 30 in. long has a diameter of ⅛ in. How many pins having lengths of 3⁄16″ can be sheared from the rod?

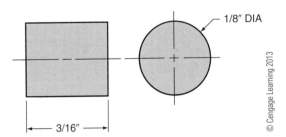

1/8″ DIA

3/16″

© Cengage Learning 2013

22. Screw threads are classified by the number of threads per inch. The pitch of a screw thread is expressed as a fraction whose numerator is 1 and whose denominator is the number of threads per inch. This bolt has a pitch of ¹⁄16. Determine the number of threads on the bolt if its length is ⅞ inches.

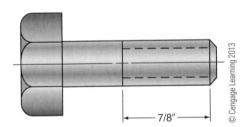

7/8″

© Cengage Learning 2013

23. Equal divisions in inches are drawn on this displacement diagram for a cam layout. Determine dimension **A.**

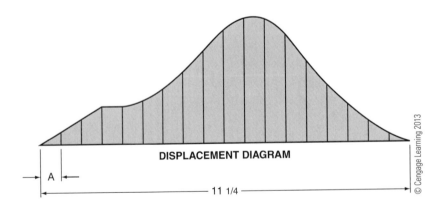

DISPLACEMENT DIAGRAM

A

11 1/4

© Cengage Learning 2013

24. A strip of 7/16-inch metal is 161 inches long. How many pieces, each
    7/8-inch long, can be stamped from it?   _____

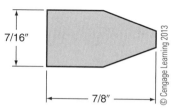

25. An isometric drawing is a three-dimensional representation of an
    object. This isometric CAD drawing must be scaled down by a factor of
    6 to make room for additional views. Its present height is 2¼″, its width
    is 10⅛″, and its depth is 8¼″. What are the new spatial dimensions in
    inches after scaling?

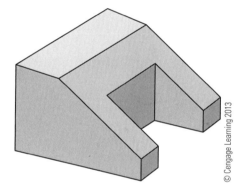

Height _____

Width _____

Depth _____

26. The CAD drawing of the arm below is to be scaled down by a factor of 2. Determine the new dimensions of **A** and **B** after scaling.

A _____

B _____

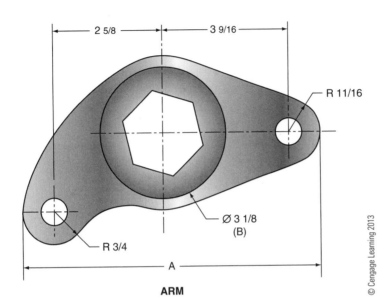

ARM

27. Use the CAD drawing below and determine the new dimensions of **A** and **B** if the drawing is scaled down by a factor of 5.

A _____

B _____

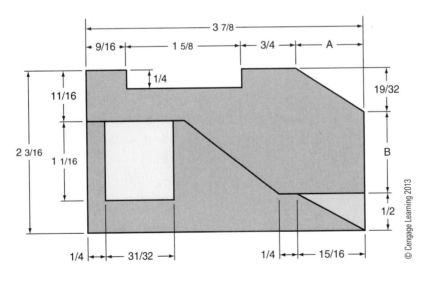

28. A CAD drafter reduces the dimensions of the arm in the CAD drawing
below by a factor of 4. What are the new dimensions of diameters **A**, **B**,
C, and **D** in the drawing?

A _____

B _____

C _____

D _____

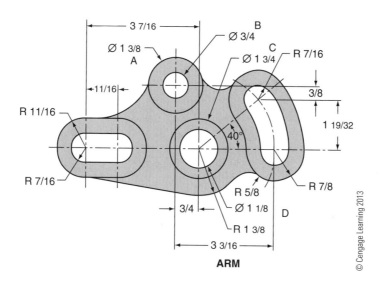

ARM

© Cengage Learning 2013

29. A CAD operator inserts the CAD drawing of a plate as a symbol.
During the insertion process, the *X*-axis (**A**) is reduced by a factor of 3,
and the *Y*-axis (**B**) is reduced by a factor of 2. Calculate the new values
of dimensions **A** and **B**.

A _____

B _____

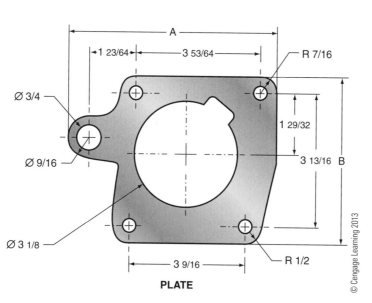

PLATE

© Cengage Learning 2013

30. The CAD drawing of the marine gasket below must be reduced to ⅓ its current size. Determine the reduced dimensions of features **A** through **D**.

A _____

B _____

C _____

D _____

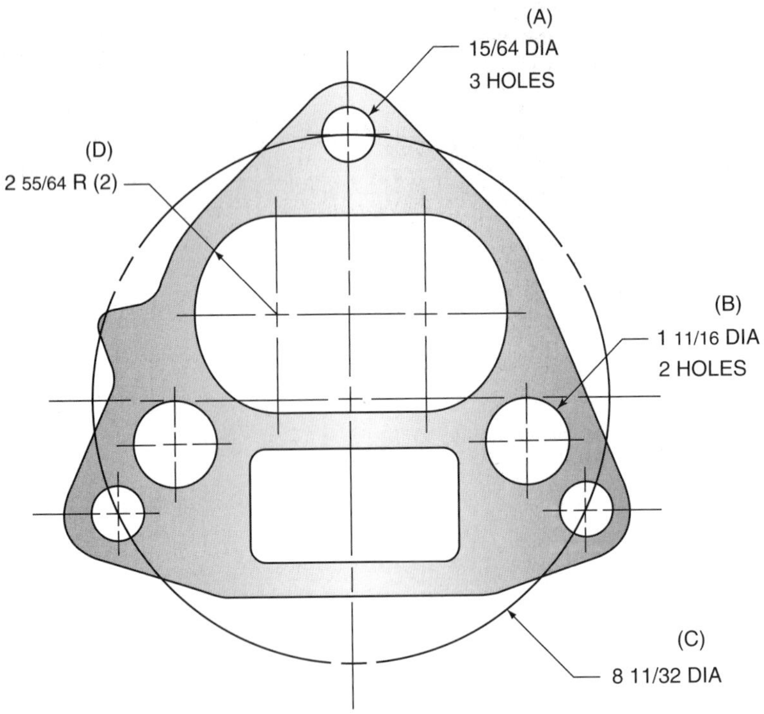

(A)
15/64 DIA
3 HOLES

(D)
2 55/64 R (2)

(B)
1 11/16 DIA
2 HOLES

(C)
8 11/32 DIA

MARINE GASKET

UNIT 10

Combined Operations

Basic Principles

This unit includes a set of mixed problems involving fractions and whole numbers and combined operations of addition, subtraction, multiplication, and division.

The properties that apply to working with fractions are the same as those that apply to working with whole numbers. Use these properties to make calculations easier.

	Addition	Multiplication
The Commutative Property	$\dfrac{1}{2} + \dfrac{2}{3} = \dfrac{2}{3} + \dfrac{1}{2}$	$\dfrac{1}{2} \times \dfrac{2}{3} = \dfrac{2}{3} \times \dfrac{1}{2}$
The Associative Property	$\left(\dfrac{1}{2} + \dfrac{2}{3}\right) + \dfrac{3}{4} = \dfrac{1}{2} + \left(\dfrac{2}{3} + \dfrac{3}{4}\right)$	$\left(\dfrac{1}{2} \times \dfrac{2}{3}\right) \times \dfrac{3}{4} = \dfrac{1}{2} \times \left(\dfrac{2}{3} \times \dfrac{3}{4}\right)$

The Distributive Property

$$3 \times \left(\frac{5}{3} + \frac{15}{16}\right) = 3 \times \frac{5}{3} + 3 \times \frac{15}{16}$$

$$3 \times \frac{5}{3} + 3 \times \frac{15}{16} = 3\left(\frac{5}{3} + \frac{15}{16}\right)$$

RULE: When simplifying expressions that have three or more operations, apply the following *order of operations.*

- Proceed from left to right and complete all multiplications and divisions in the order that they appear.

$$2 + \frac{1}{3} \times 15 - \frac{1}{2} = 2 + 5 - \frac{1}{2} \qquad \text{First multiply } \tfrac{1}{3} \text{ and 15.}$$

- Proceed from left to right again and complete all additions and subtractions in the order that they appear.

$$2 + 5 - \frac{1}{2} = 7 - \frac{1}{2} = 6\frac{1}{2}$$ Next add 2 and 5. Then subtract ½ from the sum.

Skill Problems

Simplify each expression. Include units in your answers as necessary.

1. $(\frac{3}{8}'' + 1\frac{15}{16}'') \times \frac{7}{8}$ _____

2. $(\frac{5}{6} \text{ hr.} - \frac{1}{3} \text{ hr.}) \times 8$ _____

3. $3\frac{1}{4}$ pounds $\div$ $\frac{3}{4}$ pounds $+ 6\frac{5}{8}$ pounds   _____

4. $36\frac{1}{2}$ meters $- 17\frac{3}{4}$ meters $+ 11\frac{1}{4}$ meters   _____

5. $3\frac{3}{4} \times 3\frac{1}{3} \div 2\frac{1}{2}$ _____

Practical Problems

6. What is the inch-difference in thickness between 10 pieces of ¼" brass and 6 pieces of $\frac{7}{16}$" aluminum?   _____

7. A block of steel is $1\frac{15}{16}$" thick. A rough cut of $\frac{3}{16}$" and a finish cut of $\frac{11}{64}$" are taken. What is the thickness of the remaining piece of steel in inches?   _____

8. During one week, a drafter records the following number of hours he spent on a certain project each day: Monday, 8; Tuesday, $6\frac{1}{2}$; Wednesday, $7\frac{1}{4}$; Thursday, $5\frac{2}{3}$; and Friday, $8\frac{1}{3}$. How many hours did he have to work to reach a total of 40 hours per week?   _____

9. A drafter can stamp 10 erasing shields in $7\frac{1}{2}$ seconds. At the same rate, how many erasing shields can he stamp in one minute?   _____

10. A stock rack contains four pieces of drill rod having lengths of 4½",
3³⁄₁₆", 7⅛", and 5¹³⁄₁₆". Determine the length in inches of a fifth piece so
that the length of the rack will be 26²³⁄₃₂".

                _____

11. On a scale drawing, ¼" represents 1". A line on the drawing is 9¾" long.
How many feet does the line represent?

                _____

12. Cylindrical handles are cut from handle stock. The available lengths
are 38½", 49½", 93½", and 60½". What is the total number of 5½"
handles that can be made using all this handle stock?

                _____

13. The threaded length of a threaded fastener that is less than 6 inches
long is usually 2 times the diameter of the fastener plus ¼". Calculate
the threaded length in inches of a 2⅝" bolt with a ⅝" diameter.

                _____

14. Seven holes are equally spaced apart with ¹³⁄₁₆" between the center of
each hole. Determine the distance in inches from the center of the first
hole to the center of the fifth hole.

                _____

CAD Problems

15. Determine the inside diameter (dimension **D**) of this CAD drawing of
the cross-section of pipe.

                _____

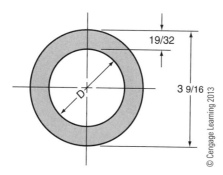

19/32

3 9/16

D

© Cengage Learning 2013

16. On this stretchout, the widths of the seven sections are equal. Dimension
     B is one-half dimension **A**. Determine dimensions **A** and **B**.   A _____

     B _____

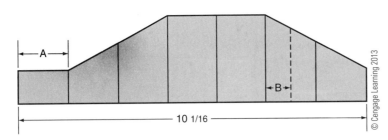

© Cengage Learning 2013

17. Dimension **A** on this drawing of a strap is ⁹⁄₁₆″ and dimension **B** is
     1³⁄₃₂″. What is **C** in inches?   C _____

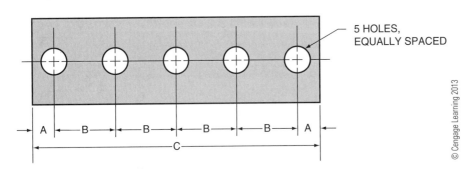

5 HOLES,
EQUALLY SPACED

© Cengage Learning 2013

18. Determine dimensions **A, B,** and **C** on this CAD drawing of a gasket.

     A _____

     B _____

     C _____

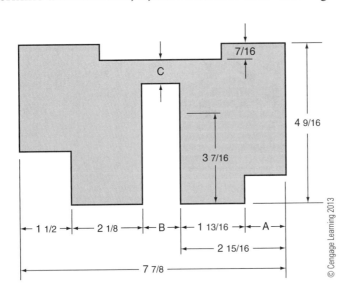

© Cengage Learning 2013

19. The sides of the 5-sided (pentagonal) cutout on the shim are each 7¼″. The sides of the 8-sided (octagonal) bar stock are each 4⅜″. How many inches greater is the distance around the pentagonal figure than around the octagonal figure?

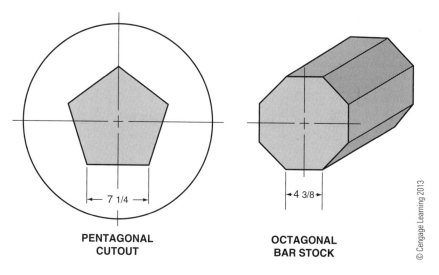

— 7 1/4 —

PENTAGONAL CUTOUT

◄4 3/8►

OCTAGONAL BAR STOCK

© Cengage Learning 2013

20. The holes on this drawing of a template are equally spaced. Calculate the overall length of the template.

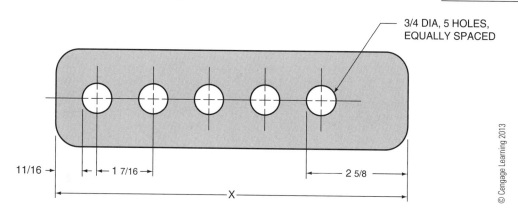

3/4 DIA, 5 HOLES, EQUALLY SPACED

11/16 →   ◄ —1 7/16—►   ◄— 2 5/8 —►

—X—

© Cengage Learning 2013

21. Use the CAD drawing of the vacation cabin below and its dimensions to determine the perimeters around the cabin and the deck. Then calculate the difference between the perimeters. Express each answer in feet and inches.

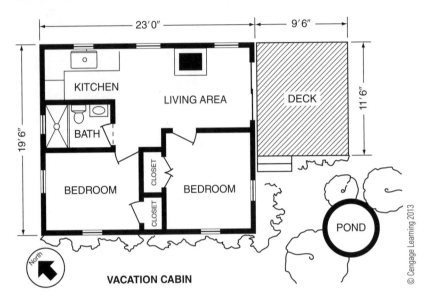

VACATION CABIN

Cabin perimeter _____

Deck perimeter _____

Difference of perimeters _____

22. The CAD drawing on the right below was scaled down from an original drawing (not shown) by a factor of 5. The dimensions of this scaled drawing are included in the drawing, and a copy of this drawing, without dimensions, is on the left.

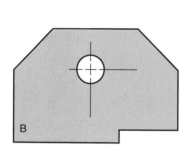

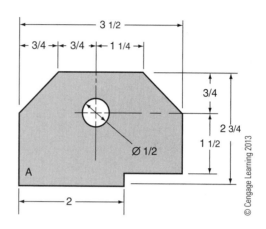

© Cengage Learning 2013

a. What was the height and width of the original drawing before it was scaled down by a factor of 5?

a. _____

b. If the copy on the left is scaled up by a factor of 9 times the diameter of the hole, what is the displacement distance from point **A** to point **B**?

b. _____

23. Use the CAD drawing below to calculate, in inches, its overall height and width, and its dimensions **A** and **B**.

Height _____

Width _____

A _____

B _____

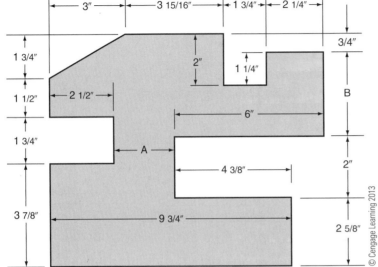

24. a. Determine dimension **A** on the CAD drawing below that includes two different size circles.

b. Determine the radii of the two unequal circles if the drawing is scaled up by a factor of 5.

Dimension A _____

R (larger circle) _____

R (smaller circle) _____

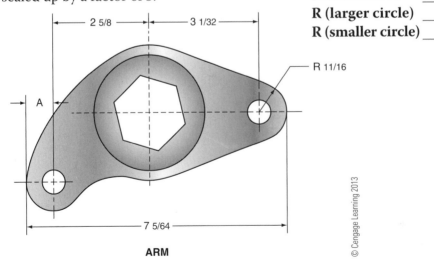

ARM

25. What are the measures of dimensions **A** through **E** in these CAD drawings?

A _____

B _____

C _____

D _____

E _____

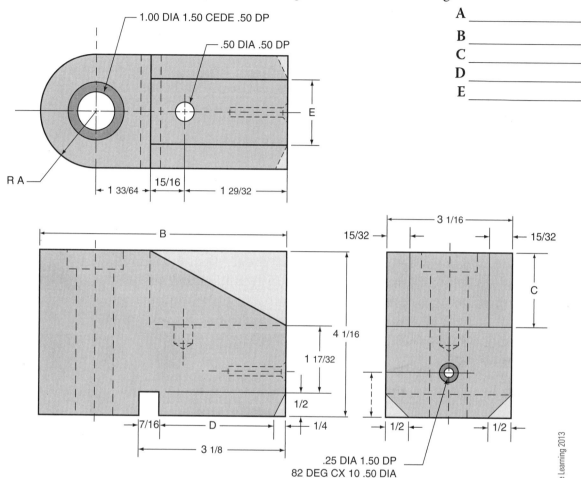

LOCATING FIXTURE

© Cengage Learning 2013

26. In the diagram below, **A** is ¹¹⁄₁₆′, **B** is 1⅜′, **C** is 2²⁹⁄₃₂′, and **E** is 3⁷⁄₆₄′.

    a. What are the dimensions of **D** and **F** in feet?

    b. What is dimension **Y** in feet if **X** is ³³⁄₆₄′, and **W** is ²⁹⁄₆₄′?

    D _____

    F _____

    Y _____

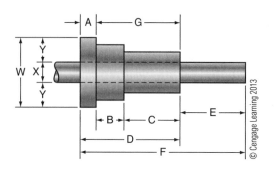

© Cengage Learning 2013

27. Calculate dimensions **A** through **D** using the CAD drawing and dimensions below. Express your answers in feet and inches. Then calculate the *inside* perimeter of the combined living room and dining areas. Disregard the two doors, archway, and fireplace in these rooms, but include the room divider in your calculations.

    A _____

    B _____

    C _____

    D _____

    Inside perimeter _____

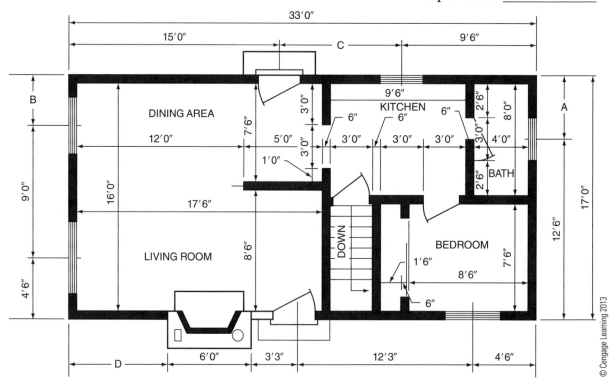

© Cengage Learning 2013

28. Use the CAD drawings below to determine dimensions **A, B,** and **C.**
What would be the length of the radius of the larger circle if the top
view were scaled up by a factor of 8?

A _____

B _____

C _____

Radius _____

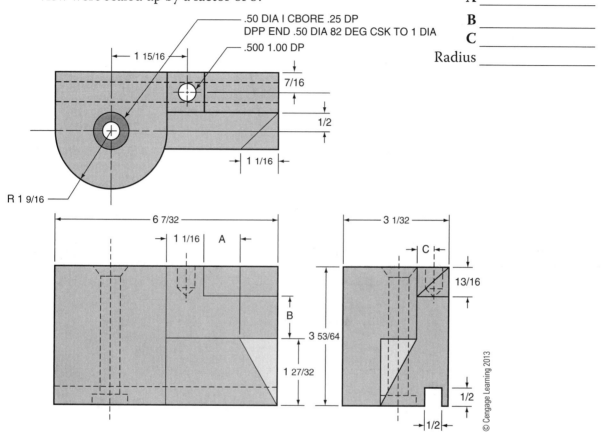

.50 DIA I CBORE .25 DP
DPP END .50 DIA 82 DEG CSK TO 1 DIA

.500 1.00 DP

1 15/16

7/16

1/2

1 1/16

R 1 9/16

6 7/32

1 1/16   A

B

3 53/64

1 27/32

3 1/32

C

13/16

1/2

1/2

© Cengage Learning 2013

29. Calculate dimension **A** using the CAD drawing below. Also determine
the lengths of the diameters of circles (1) and (2) if the object is scaled
up by a factor of 5.

A _____

Diameter (1) _____

Diameter (2) _____

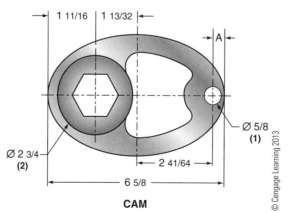

1 11/16   1 13/32

A

Ø 5/8
(1)

Ø 2 3/4
(2)

2 41/64

6 5/8

CAM

© Cengage Learning 2013

30. Use the CAD drawing of the shim below and calculate the difference between the outer perimeter of the shim and the sum of the perimeters of the inner hexagon and the two square cut-outs. The shim is both horizontally and vertically symmetrical.

Outer perimeter _____

Inner perimeters (sum) _____

Difference _____

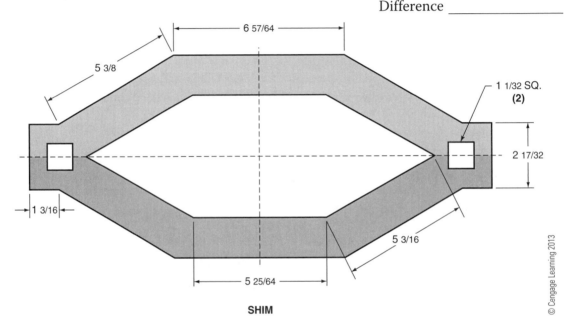

SHIM

31. Use the CAD drawing of the reference gauge below and calculate
    dimensions **A** through **E.**

A _____

B _____

C _____

D _____

E _____

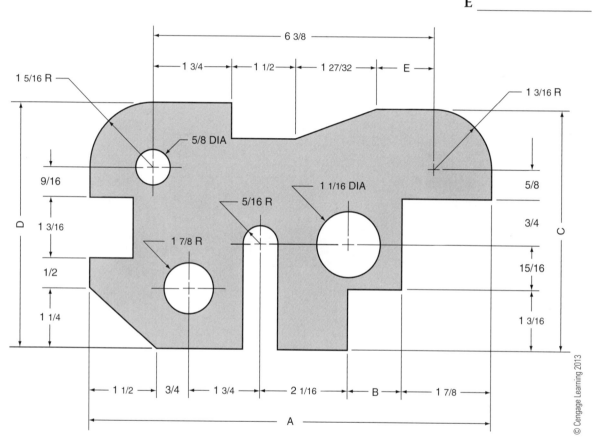

© Cengage Learning 2013

32. Determine the values for dimensions **A** through **F** in the CAD drawing
of the gasket below.

A _____
B _____
C _____
D _____
E _____
F _____

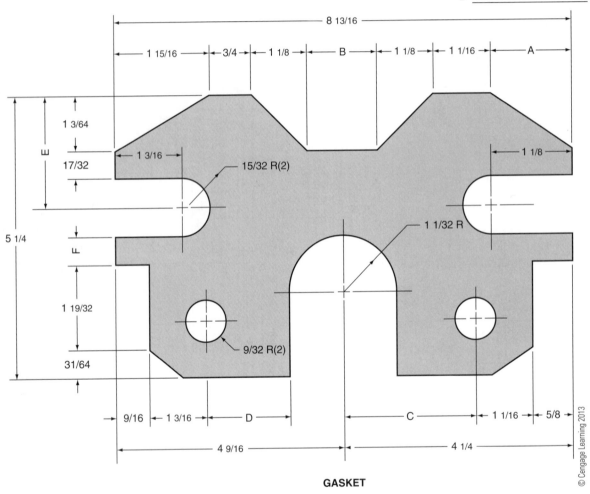

GASKET

SECTION

Decimals

UNIT 11

Addition

Basic Principles

A decimal, or *decimal fraction* as it is sometimes called, is a fraction whose denominator is a power of 10. Examples include 1/10, 1/100, 1/1000, and so on. The power, or exponent, of 10 indicates the number of zeroes that follow the digit 1.

$10^1 = 10$	10 to the *first* power
$10^2 = 100$	10 to the *second* power, or 10 squared
$10^3 = 1,000$	10 to the *third* power, or 10 cubed
$10^4 = 10,000$	10 to the *fourth* power
$10^n = \underbrace{1000...00}_{n \text{ zeroes}}$	10 to the nth power

NOTE: 10^{100} is the digit 1 followed by 100 zeroes. This very large number is called a *googol*.

In drafting applications and measurements, the numerator of a decimal fraction is usually a positive whole number.

EXAMPLES: $\dfrac{5}{10}, \dfrac{9}{100}, \dfrac{17}{1000}, \dfrac{255}{10,000}$

A decimal consists of two parts: a positive or negative whole number, or 0, and a decimal number. A decimal point (.) separates the two parts. You can also write the decimal part of the number as an equivalent fraction whose denominator is a power of 10.

EXAMPLE: 213.605, where $213.605 = 213\dfrac{605}{1000}$

Whole number part   Decimal part   Decimal fraction

You can use a place value chart to identify the place value of the digits in a decimal number, such as 213.605. (Read this number as "two hundred thirteen and six hundred five thousandths".)

hundreds	tens	ones		tenths	hundredths	thousandths
2	1	3	.	6	0	5

When a decimal is written as a decimal fraction (or as the decimal fraction in a mixed number), the denominator is the power of 10 in the farthest place to the right of the decimal point.

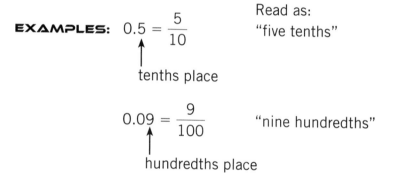

Read as:

EXAMPLES: $0.5 = \dfrac{5}{10}$ "five tenths"

↑
tenths place

$0.09 = \dfrac{9}{100}$ "nine hundredths"

↑
hundredths place

$2.045 = 2\dfrac{45}{1000}$ "two and forty five thousandths"

↑
thousandths place

NOTE: Read the decimal point as the word "and."

To add decimals line up the decimal points in each number and the corresponding digits in each place. If necessary, insert zeroes after the last digit to the right of the decimal point in each decimal so there are digits in every place of the numbers. Then add the digits in each place, starting at the farthest place on the right. Use regrouping if the sum in any place is greater than 9.

EXAMPLE: $0.7 + 2.6 + 141.756 + 36$

Insert trailing zeroes after 0.7, 2.6, and 36, so that there are three decimal places in each addend.

$$
\begin{array}{r}
\overset{1\ 2}{0.700} \\
2.600 \\
141.756 \\
+\ 36.000 \\
\hline
181.056
\end{array}
$$

The sum is $181.056 = 181\dfrac{56}{1000}$.

📟 $0.7 \boxplus 2.6 \boxplus 141.756 \boxplus 36 \boxed{=}$

NOTE: Press the $\boxed{\cdot}$ key on the calculator keypad to enter a decimal point.

Skill Problems

Calculate the sums.

1.   0.75
      + 0.137

2.   0.245
      + 0.767

3.   0.5018
      + 0.3754

4.   0.987
      + 0.335

5. $0.7325 + 0.112 + 0.416$ _____

6. $2.1307 + 3.718 + 4.86$ _____

7. $21.601 + 7.009 + 16.736$ _____

8. $0.017 + 4.25 + 14.379 + 8.3$ _____

Practical Problems

9. Calculate the total cost of the following drafting instruments: a 6-inch bow compass, $9.89; a mechanical drafter's scale, $9.95; a drafter's mechanical pencil, $3.98; and an 8-inch adjustable triangle, $29.59. _____

10. The thicknesses of six machine parts are measured with a micrometer. The measurements are 0.065", 0.145", 0.7146", 1.606", 0.329", and 3.250". What is the total thickness, in inches, of the parts? _____

11. During each of the five days in one week, a drafter worked 8.75 hours, 10.25 hours, 6.5 hours, 8 hours, and 7.25 hours. How many total hours did the drafter work that week?

_____

12. A file clerk's weekly paycheck had the following deductions: $8.10 for medical insurance, $56.13 for federal income tax, $12.17 for FICA, $4.10 for state income tax, and $2.35 for union dues. Compute the total amount of deductions.

_____

13. A mechanical drafter measures off segments on a center line. The lengths she measures are 3.75 centimeters, 6.25 centimeters, 0.60 centimeters, 7.37 centimeters, and 5.60 centimeters. What is the overall length, in centimeters, of all the segments on the center line?

_____

14. During a 2-month period, a drafter bought various drafting supplies that cost $37.50, $18.75, $21.60, $45.80, $27.33, and $12.88. How much did all of these supplies cost?

_____

15. A CAD drafter has devoted several weeks to a structural steel project. The hours he spent on various aspects of the job were 40.05, 37.25, 146.75, 0.40, 92.45, and 112.15. Calculate the total number of hours the CAD drafter worked on this project.

_____

16. A CAD drafter purchased a CAD station for use at home. The computer and software cost a total of $5,675.00. Later he added a plotter that cost $1,997.00, a laser printer that cost $169.99, and an external hard drive that cost $149.99. What was the total cost of his equipment?

_____

CAD Problems

17. What is the overall length of this link?

_____

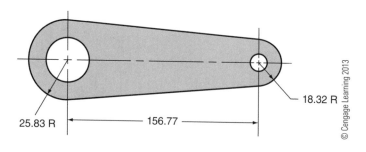

25.83 R   156.77   18.32 R

© Cengage Learning 2013

18. The CAD drawing below shows dimensions on a gasket. Determine the perimeter of this gasket.

_____

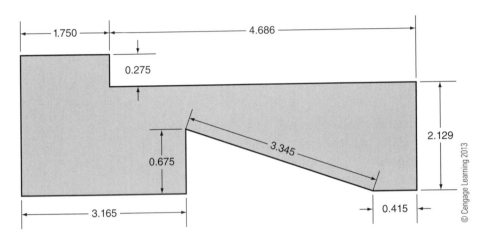

19. Calculate the overall length on this CAD drawing of a sliding lever.

_____

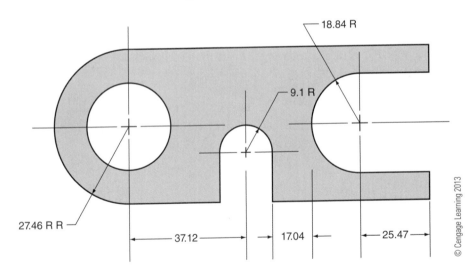

20. Determine dimensions **A** and **B** on this CAD drawing of a strap.

A _____

B _____

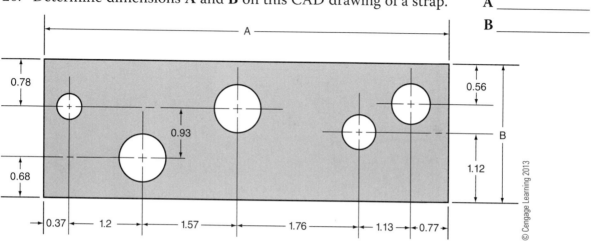

21. The figure below is an isometric CAD drawing of a step block. Determine dimension A on this step block.

_____

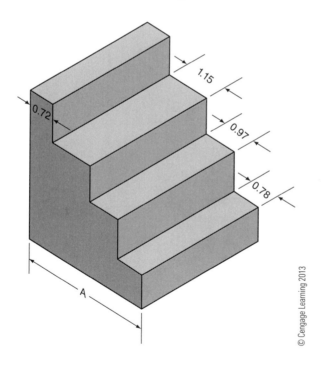

© Cengage Learning 2013

22. Determine the height and width in inches of the CAD drawing below.

Height _____

Width _____

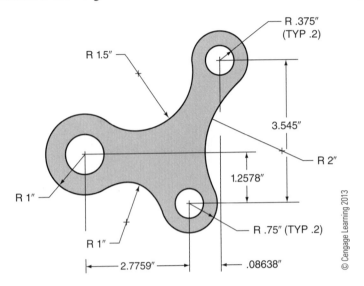

R .375"
(TYP .2)

R 1.5"

3.545"

R 2"

1.2578"

R 1"

R .75" (TYP .2)

R 1"

2.7759"

.08638"

© Cengage Learning 2013

23. Use the CAD drawing of the plot plan below and calculate its perimeter in feet.

_____

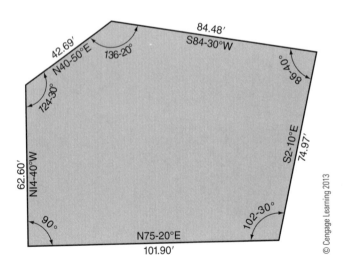

84.48'
S84-30°W

42.69'
N40-50°E

136-20°

124-30°

N4°-40°

98-40°0°

62.60'
NI4-40°W

S2-10°E
74.97'

90°

102-30°

N75-20°E
101.90'

© Cengage Learning 2013

24. Determine the overall length of the CAD drawing of the cam below. _____

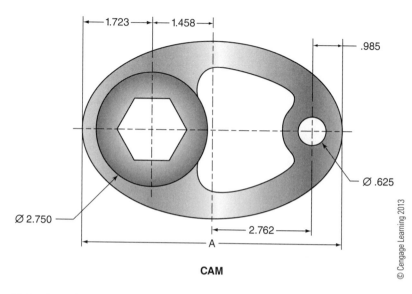

CAM

25. Calculate dimensions **A, B,** and **C** in the CAD drawings below.

A _____

B _____

C _____

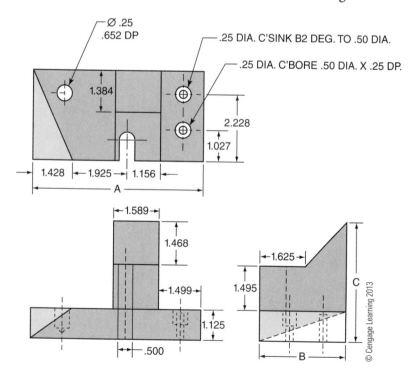

26. Use the CAD drawing of the arm below to calculate dimension A. _____

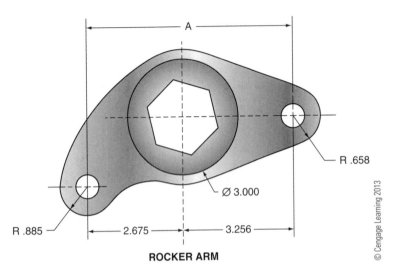

R .658

Ø 3.000

R .885

2.675

3.256

ROCKER ARM

© Cengage Learning 2013

27. Determine dimensions **A, B,** and **C** in the CAD drawing of the dovetail slide below.

A _____

B _____

C _____

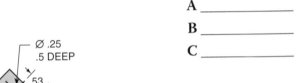

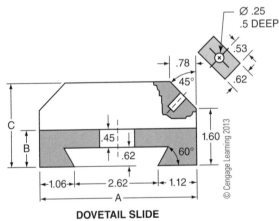

Ø .25
.5 DEEP

.53

.78

45°

.62

C

.45

1.60

B

.62

60°

-1.06-

2.62

1.12

A

DOVETAIL SLIDE

© Cengage Learning 2013

28. Use the CAD drawing below to calculate its overall height and width in inches.

Height _____

Width _____

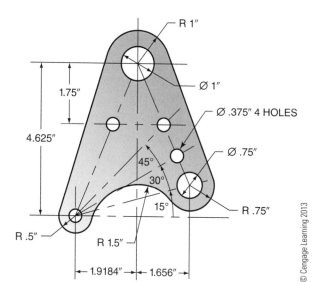

29. Use the CAD drawing of the plate below to determine dimensions **A** and **B**.

A _____

B _____

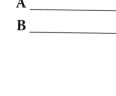

PLATE

30. Use the CAD drawing of a gasket below to determine dimensions **A** through **E**.

A _____

B _____

C _____

D _____

E _____

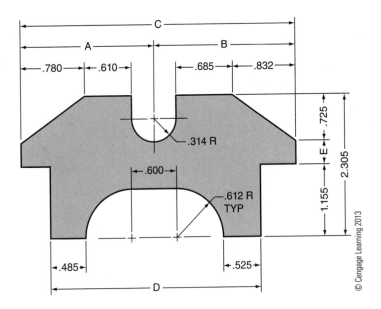

31. Determine the outside and inside perimeters of the CAD drawing below.

Inside perimeter _____

Outside perimeter _____

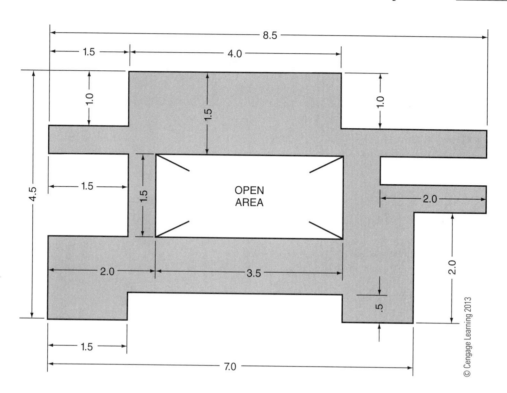

© Cengage Learning 2013

UNIT 12

Subtraction

Basic Principles

To subtract decimal numbers, write the lesser number under the greater number. Be sure to align the decimal points in the numbers and the corresponding digits in each place.

EXAMPLE: $19.643 - 8.132$

$$
\begin{array}{r}
19.643 \\
- \ 8.132 \\
\hline
11.511
\end{array}
$$

The difference is $11.511 = 11^{511}/_{1000}$.

 $19.643 \ \boxed{-} \ 8.132 \ \boxed{=}$

Skill Problems

Calculate the differences.

1. $\begin{array}{r} 0.87 \\ -\ 0.28 \\ \hline \end{array}$

2. $\begin{array}{r} 0.349 \\ -\ 0.063 \\ \hline \end{array}$

3. $\begin{array}{r} 1.725 \\ -\ 0.659 \\ \hline \end{array}$

4. $\begin{array}{r} 8.375 \\ -\ 3.683 \\ \hline \end{array}$

5. $18.6 - 10.89$ _____

6. $21.634 - 9.397$ _____

7. $118.717 - 77.9$ _____

8. $675.6 - 486.857$ _____

Practical Problems

9. The outside diameter of a circular shaft tapers from 2.8750" to 2192". What is the difference in inches between these diameters?

10. A block of steel is 25.40 mm thick. One cut of 3.25 mm and a second cut of 4.17 mm are taken. What is the remaining thickness of the block in millimeters?

11. A civil drafter has a $725 budget for equipment. She buys a new drafting machine that costs $406, a machine scale that costs $31.99, and an electronic calculator that costs $117.86. How much money does the drafter have left in her budget?

12. Two reams of paper weigh 3.87 pounds and 14.73 pounds, respectively. Calculate, in pounds, the difference between the weights.

13. An architectural drafter's bill for new equipment is $167.74. The discount offered on the purchase is $16.75. How much does the new equipment cost?

14. A steel block is 79.315 cm long. A machinist cuts off 6.25 cm. What is the length in centimeters of the block after machining?

15. The outside diameter (OD) of a drilled hole is 0.7812". The hole is made larger with a reamer having an OD of 0.7969". Determine, in inches, the increase in length of the diameter of the hole after the reaming operation.

16. A clearance fit of 0.012 inches is needed between a shaft and a bushing to work properly. The inside diameter (ID) of the bushing measures 1.475 inches. What must be the measure in inches of the outside diameter (OD) of the shaft to obtain this clearance?

17. A CAD operator makes the following purchases: $1,180.00 for CAD software, $52.95 for disk management software, $99.95 for a memory upgrade kit, and $599.00 for a paint program. A $250 cash rebate was given at the time of purchase. How much did the CAD operator pay for these products?

CAD Problems

18. What is the length of the threaded portion in this CAD drawing of a round-head cap screw? _____

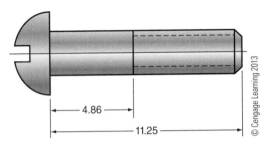

4.86

11.25

19. Two pins are to be located in this CAD drawing of a steel base, and the distance **A** between the centers of the pins must be determined. Determine dimension **A**. _____

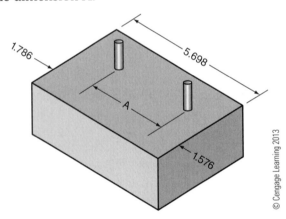

1.786

5.698

A

1.576

20. The length of the front view in the CAD drawing on the left below is 257.37 mm. The width of its right profile view is 147.82 mm. Calculate, in millimeters, the difference between these measurements. _____

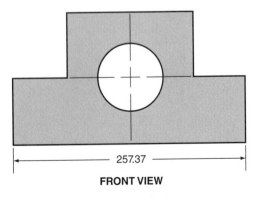

257.37

FRONT VIEW

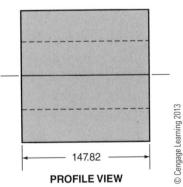

147.82

PROFILE VIEW

21. Determine dimensions **A** and **B** in this CAD diagram of a shim.

A _____

B _____

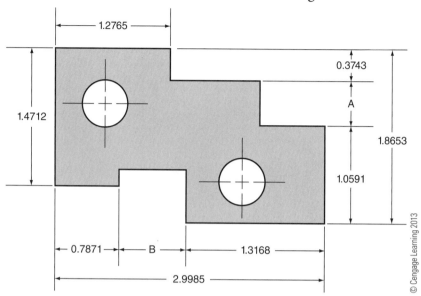

22. Use the CAD drawing below of a rocker arm and determine dimension **A**.

_____

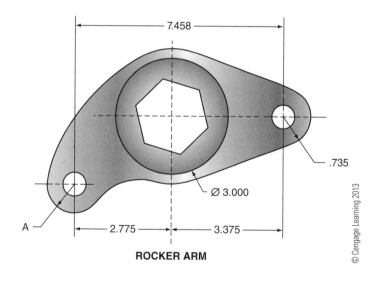

ROCKER ARM

© Cengage Learning 2013

23. Use the CAD drawing below and determine dimensions **A, B,** and **C.**

A _____

B _____

C _____

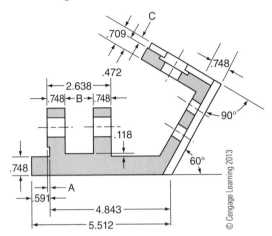

24. The perimeter of the CAD drawing below is 22.3297. Calculate the length of line **AB.**

_____

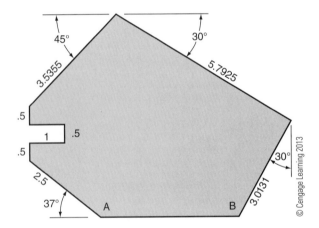

25. Use the CAD drawing below to calculate the difference between the outside perimeter and the sum of the perimeters of the inside features. _____

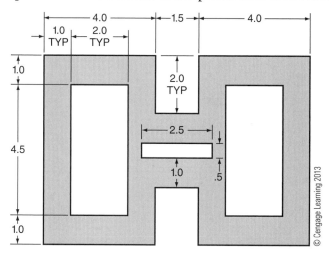

26. Determine dimension **A** in the CAD drawing of the cam below. _____

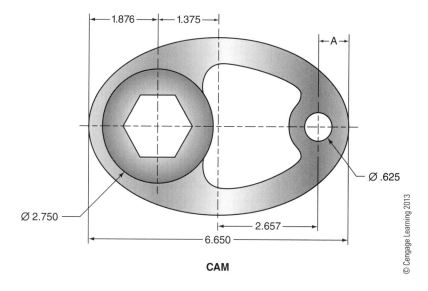

CAM

27. The perimeter of the plot in the CAD drawing below is 2812.9'.
    Calculate in feet the distance between points **E** and **F**.   _____

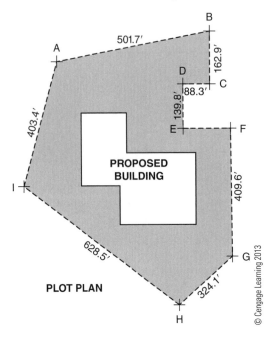

PLOT PLAN

28. Use the CAD drawing of the symmetrical shim below to calculate **A**,
    the outer perimeter of the shim; **B**, the sum of the perimeters of the
    inner features; and **C**, the difference between the two perimeters.

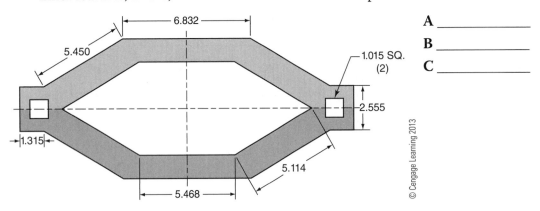

A _____

B _____

C _____

29. Determine dimensions **A** through **E** on the top and front views of these CAD drawings of a retainer.

A _____

B _____

C _____

D _____

E _____

RETAINER

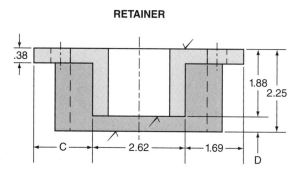

TOP VIEW

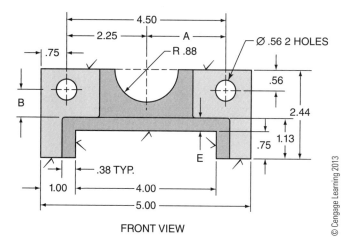

FRONT VIEW

© Cengage Learning 2013

30. Use the CAD drawing below to calculate dimensions **A** through **D**.

A _____
B _____
C _____
D _____

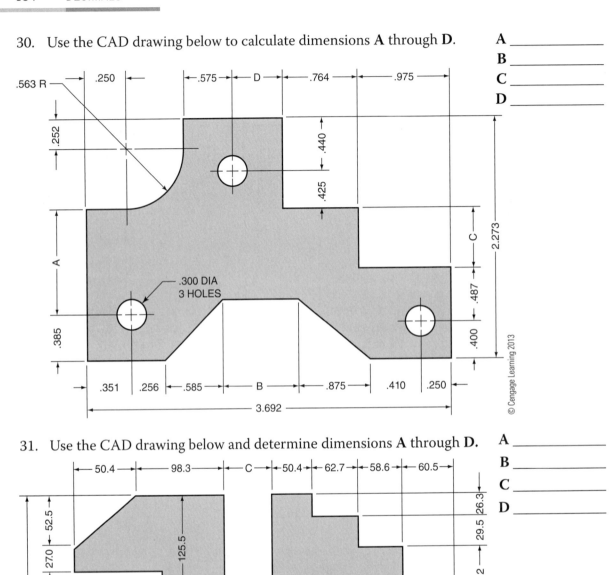

31. Use the CAD drawing below and determine dimensions **A** through **D**.

A _____
B _____
C _____
D _____

© Cengage Learning 2013

UNIT 13

Multiplication

Basic Principles

Decimal numbers are multiplied in the same way as whole numbers. To determine the number of decimal places in a product, add the number of decimal places in each factor. Then start at the far right of the whole-number product and move the decimal point that many places to the left.

EXAMPLE: 2.3 × 12.15

Whole numbers	Decimal numbers	
1215	12.15	← 2 decimal places
×23	×2.3	← 1 decimal places
3645	3645	
+2430	+2430	
27945	27.945	← 3 decimal places

The decimal point in the decimal product is moved three places to the left of the digit 5 in the ones place of the whole-number product 27,945.

 2.3 ☒ 12.15 ▣

Rounding Decimals

It is often useful or necessary to *round* a decimal number to a specified place. To round a number, look at the place immediately to the right of the place to be rounded.

- If the digit is greater than or equal to 5, round the digit in the rounding place up to the next higher digit.

135

EXAMPLE: Round 25.369 to the hundredths place.

The digit 9 is in the thousandths place, and 9 > 5. Therefore, drop 9 and round 6 in the hundredths place up to 7.

$$25.369 \rightarrow 25.37$$

- If the digit to the immediate right is less than 5, drop the digit(s) to the right of the rounding place.

EXAMPLE: Round 4.639 to the tenths place.

The digit 3 is in the hundredths place, and 3 < 5. Therefore, drop all the digits to the right of the tenths place.

$$4.639 \rightarrow 4.6$$

Skill Problems

Calculate the products.

1. 0.83
    $\times$ 0.06

2. 0.78
    $\times$ 0.23

3. 0.465
    $\times$ 0.87

4. 1.714
    $\times$ 0.32

5. 13.64 $\times$ 0.27   _____

6. 27.321 $\times$ 0.73   _____

7. 32.18 $\times$ 0.69   _____

8. 43.76 $\times$ 2.15   _____

Practical Problems

9. A bolt has a diameter of 1.375 inches. The threaded length of the bolt is 2 times the diameter plus 0.250 inch. What is the threaded length of the bolt in inches?   _____

10. A designer knows that for each complete turn of a wheel, the wheel rolls a distance approximately equal to 3.14(π) times its diameter. How far in inches does a 12.6 inch-diameter wheel roll after three complete turns?   _____

11. A certain type of Mylar film costs $4.50 a sheet. What is the cost of 144 sheets?

    _____

12. A plastic drafter's scale weighs 0.632 pounds. What is the weight in pounds of 48 scales?

    _____

13. A CAD drafter's salary is increased by $63.28 per month. What is the drafter's total raise for the year?

    _____

14. A 3D projection of a circle is called an *isometric ellipse*. The isometric ellipse shown below is 1.225 times larger than the circle it represents. The diameter of the circle is 0.750". What is the length in inches of the longer axis of the ellipse?

    _____

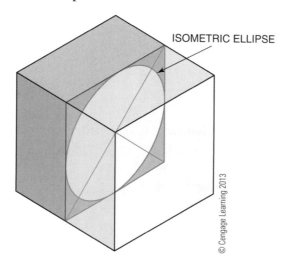

ISOMETRIC ELLIPSE

© Cengage Learning 2013

15. A microfilmed drawing is increased by a factor of 2.75. A line on the drawing is 5.25 cm long. What is the new length, in centimeters, of the line?

    _____

16. A CAD drafter drives 45.5 miles to work each day. What is his total round-trip mileage if the drafter works 165.5 days a year, and the distances each day remain the same?

    _____

17. The cost of a wireless computer mouse is $19.95. What is the cost of six of this type of mouse?

    _____

18. A USB flash drive costs $9.99. Calculate the cost of three drives. What would be the cost of a gross of drives? (A gross contains 12 dozen or 144 items.)

Dozen _____

Gross _____

CAD Problems

19. A shaft is tapered 0.375″ per inch of its length. By how many inches is the CAD drawing of the shaft below tapered?   _____

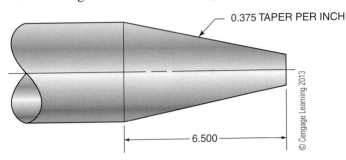

0.375 TAPER PER INCH

6.500

© Cengage Learning 2013

20. The spacing collar in the CAD drawings below is 0.375″ wide. How many inches will 19 spacing collars take up on a shaft?   _____

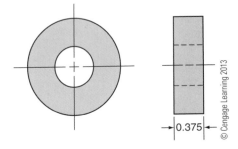

0.375

© Cengage Learning 2013

21. All sections on this CAD drawing of a gauge have equal widths. What is the overall length of the gauge?

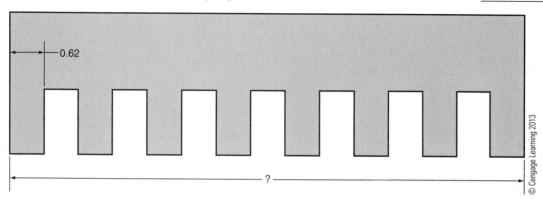

0.62

?

© Cengage Learning 2013

22. There are 17 equally spaced holes on this CAD drawing of a plate. Calculate the distance between the centers of holes **A** and **B**.

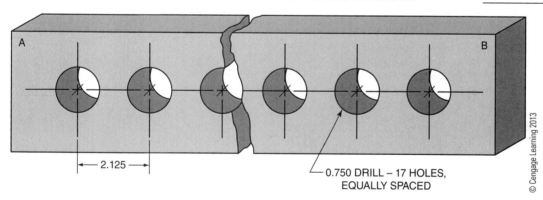

A

B

2.125

0.750 DRILL – 17 HOLES,
EQUALLY SPACED

© Cengage Learning 2013

23. This round-head cap screw has a thread pitch of ¹⁄₁₂; that is, 12 threads per inch. How many threads are contained in a rod that is 11.625 inches long and has the same thread pitch as the CAD drawing of this cap screw?

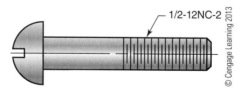

1/2-12NC-2

© Cengage Learning 2013

24. Determine the height, width, and depth after this isometric CAD drawing of an object is scaled up by a factor of 4.

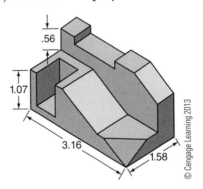

Height _____

Width _____

Depth _____

25. This pump symbol is to be inserted into a CAD drawing at 6 times its present size. Calculate the new height and width of the pump if the grid size is 0.125.

Height _____

Width _____

INSERTION
POINT

PUMP

© Cengage Learning 2013

26. In the CAD drawing below, the larger holes must be increased by a factor of 3.25, and the small hole must be increased by a factor of 2.75. Determine the sizes of these enlargements.

Large holes _____

Small hole _____

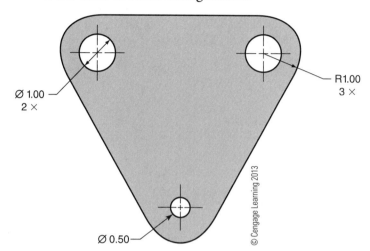

⌀ 1.00
2 ×

R1.00
3 ×

⌀ 0.50

© Cengage Learning 2013

27. The symbol for an exchanger used in piping is shown on a 0.25 grid. The symbol is inserted using a 1.75 scaling factor. Calculate the enlarged sizes for its height and width.

Height _____

Width _____

INSERTION
POINT

EXCHANGER

© Cengage Learning 2013

28. The CAD drawing below is to be scaled up by a factor of 5. Calculate the new dimensions for its length and width.

Length _____

Width _____

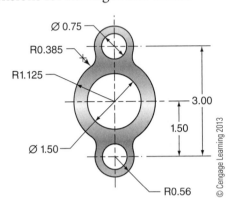

29. Determine dimension **A** and the overall width and height of the CAD drawing of the shim below.

A _____

Width _____

Height _____

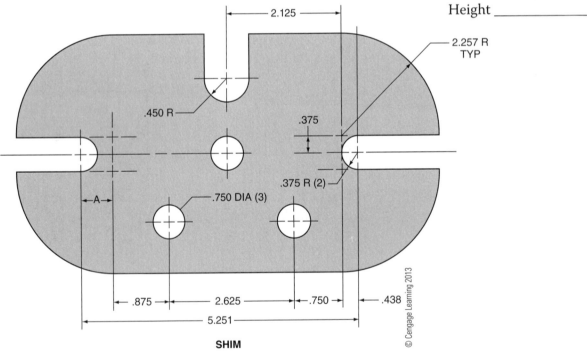

SHIM

UNIT 14

Division

Basic Principles

When dividing a number by a decimal the divisor must be rewritten as a whole number. To do this, write the problem as a fraction. Then multiply the fraction by 1 written as an improper fraction, such as $^{10}/_{10}$, $^{100}/_{100}$, and so on.

EXAMPLE: $103.84 \div 5.9$

STEP 1:   Write the division problem as a fraction. The divisor is the denominator.

$$103.84 \div 5.9 = \frac{103.84}{5.9}$$

STEP 2:   Because there is just one decimal place in the divisor, multiply the fraction by the improper fraction $^{10}/_{10}$. This results in a whole-number denominator and 1038.4 in the numerator. Now divide 1038.4 by 59.

$$\frac{103.84}{5.9} \times \frac{10}{10} = \frac{1038.4}{59}$$

NOTE: This step is equivalent to moving the decimal point in both numbers one place to the right.

STEP 3: Use long division to calculate the quotient.

```
                    17.6  ← Quotient
Divisor → 59)1038.4  ← Dividend
          −59
          448
         −413
          354
         −354
            0
```

⊞ 103.84 ÷ 5.9 =

Remember, to round a number, look at the place immediately to the right of the place to be rounded.

- If the digit is greater than or equal to 5, round the digit in the rounding place up to the next higher digit.
- If the digit to the right of the rounding place is less than 5, drop all of the digit(s) after the rounding place.

Skill Problems

Divide and express each quotient to the indicated decimal place.

1. 0.96 ÷ 4   (hundredths) _____

2. 99.19 ÷ 0.7   (tenths) _____

3. 8.75 ÷ 1.25   (tenths) _____

4. 425.86 ÷ 62   (thousandths) _____

5. 0.0057 ÷ 19   (ten thousandths) _____

6. 0.0988 ÷ 0.012   (thousandths) _____

7. 73.275 ÷ 2.5   (hundredths) _____

8. 6.111 ÷ 0.83   (tenths) _____

Practical Problems

9. The dividers in a box weigh 16.32 lb. Each set of dividers weighs 0.68 lb. How many dividers are in the box?

10. One thousand drafting triangles weigh 379.7 pounds. What is the weight of each triangle?

11. A ¼-inch diameter screw has 28 threads per inch. Calculate, to the nearest thousandth inch, the pitch of the screw. (Pitch = $1/n$, where n = the number of threads per inch.)

12. Four drafters work on a special job after regular working hours. They receive $726 to be shared equally. How much money does each drafter receive?

13. Seventeen drafters receive a total bonus of $36,321.50 for completing a project on time. To the nearest cent, if the money is shared equally, how much does each drafter receive?

14. An object line on a detail drawing is 17.509 inches long. If the line is divided into 19 equal parts, what is the length in inches of each part, rounded to the nearest thousandth?

15. A CAD drafter locates circles of diameters 0.375, 1.625, 3.250, 2.438, 4.500, 1.922, and 5.187 on a drawing. Calculate, to the nearest thousandth, the average length of these diameters. (An average is equal to the sum of the numbers divided by the number of numbers.)

16. A civil drafter receives $826.50 for a 38-hour week. Calculate the drafter's hourly rate of pay.

17. A box of twelve 4H drawing pencils costs $22.44. What is the cost of one box?

18. A CAD drafter records the following number of hours she devoted to architectural projects: 17.20, 15.75, 19.5, 16.4, 18.25, 11.0, and 14.30. What is the average number of hours she spent on these projects, rounded to the nearest hundredth?

19. A drawing board is 0.750″ thick. A stack of these boards is 27.750″ high. How many boards are in the stack?

CAD Problems

20. The pitch of a screw thread is found by dividing 1.00 inch by the number of threads per inch. A 1¼-13 ACME screw thread has 13 threads per inch. What is its thread pitch to the nearest thousandth inch? (Pitch = $1/n$, where n = the number of threads per inch) _____

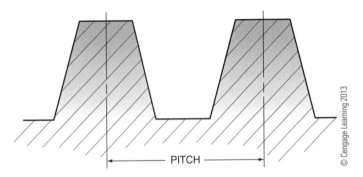

PITCH

© Cengage Learning 2013

21. All sections on this CAD development drawing are equal in length. Calculate dimension **A**. _____

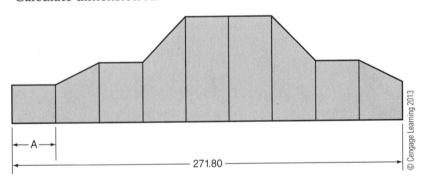

A

271.80

© Cengage Learning 2013

22. All sections on the CAD drawing of this block are equal in length. Determine length **X** to the nearest thousandth. _____

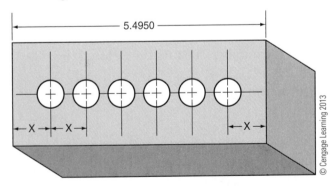

5.4950

X   X   X

© Cengage Learning 2013

23. A regular hexagon has six equal sides. If the perimeter of this regular hexagon is 250.029 mm, what is the length of each side in millimeters? _____

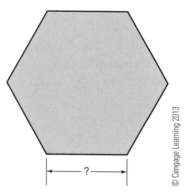

24. Use the CAD drawing below and determine the new size of the holes that must be reduced by a factor of 6 in order to work with another part.

_____

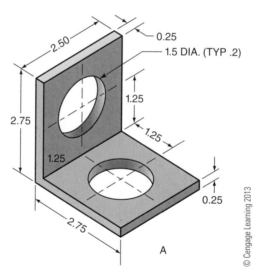

25. This drawing of a pressure safety valve is inserted in a CAD drawing with a reduction factor of 6. If the grid spacing is 0.75, calculate the new height and width of the valve.

Height _____

Width _____

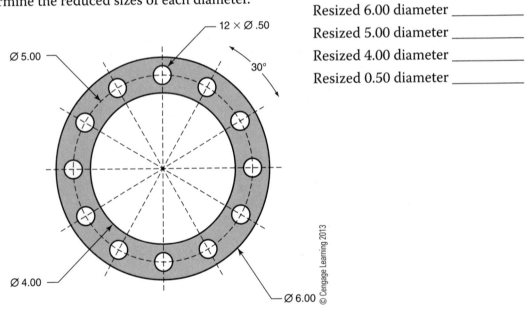

INSERTION POINT

PRESSURE SAFETY VALVE

© Cengage Learning 2013

26. The CAD drawing of the spacer ring below can be modified to fit with another part if the 6.00. 5.00. and 4.00 diameter circles are reduced by a factor of 3, and the 0.50 diameter holes are reduced by a factor of 4. Determine the reduced sizes of each diameter.

Resized 6.00 diameter _____

Resized 5.00 diameter _____

Resized 4.00 diameter _____

Resized 0.50 diameter _____

12 × Ø .50

30°

Ø 5.00

Ø 4.00

Ø 6.00

© Cengage Learning 2013

27. Use the CAD drawings of the notched block below to determine the
new values for dimensions **A** through **E** if the block is scaled down by a
factor of 3. Round the answers to the nearest thousandth.

A _____

B _____

C _____

D _____

E _____

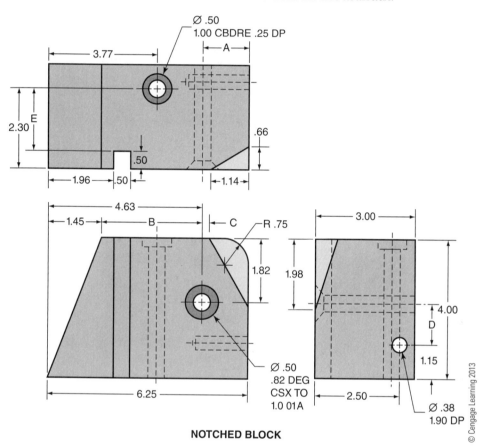

NOTCHED BLOCK

© Cengage Learning 2013

UNIT 15

Basic Principles

The five addition and multiplication properties for working with decimals and applying the order of operations when simplifying expressions are the same as those for whole numbers and fractions.

- The Commutative Properties

Addition	Multiplication
$2.4 + 1.06 = 1.06 + 2.4$	$2.4 \times 1.06 = 1.06 \times 2.4$

- The Associative Properties

Addition

$$(2.4 + 1.06) + 0.5 = 2.4 + (1.06 + 0.5)$$

Multiplication

$$(2.4 \times 1.06) \times 0.5 = 2.4 \times (1.06 \times 0.5)$$

- The Distributive Property

$$2.4 \times (1.06 + 0.5) = 2.4 \times 1.06 + 2.4 \times 0.5$$
$$\text{and}$$
$$2.4 \times 1.06 + 2.4 \times 0.5 = 2.4 \times (1.06 + 0.5)$$

- The Order of Operations
  1. Start at the far left of the expression and complete all multiplications and divisions in the order that they appear.
  2. Start at the far left again and complete all additions and subtractions in the order that they appear.

Skill Problems

Use the properties of numbers and order of operations to simplify each expression. Express each answer to the nearest thousandth as necessary.

1. $0.256 \times (1.119 - 0.430)$ _____

2. $15.065 + 0.98 \times 0.673$ _____

3. $12.67 + 32.468 \div 3.7$ _____

4. $57.95 - 114.64 \div 6.3$ _____

5. $(17.63 + 39.257) \times 4.8$ _____

6. $7.2 \times 13.5 \div 4.7$ _____

Practical Problems

7. Two holes have diameters of 2.7625" and 1.3156". Calculate the difference in inches between the radii. _____

8. Three holes have diameters of 2.187", 1.250", and 6.373". Calculate the number of inches in the sum of the three radii. _____

9. A block is 7.188" thick. Three rough cuts of 0.473" each and one finish cut of 0.012" are machined from the block. What is the remaining thickness of the block in inches? _____

10. Artgum erasers are produced at a rate of 2,880 erasers per hour. At that rate, how many erasers can be made in 3 minutes and 45 seconds? _____

11. A supervisor set up this chart to calculate the cost of drafting pencils that he needs to purchase. Complete the chart and answer the questions that follow.

Type	Cost/dozen	Number needed	Total cost/type
F	$15.00	36 dozen	
H	$10.45	15 dozen	
2H	$ 9.29	24 dozen	
6H	$12.38	18 dozen	

a. How much will it cost to buy all the pencils?   a. _____

b. How many pencils of all types will he purchase in all?   b. _____

12. Two drafting departments work on a large job. Department A works 59.5 hours, 87.8 hours, 106.7 hours, 37.4 hours, and 77.3 hours. Department B works 97.9 hours, 62.7 hours, 48.3 hours, and 102.75 hours. How many hours more does Department A work than Department B? _____

13. A small CAD department purchased the following items: a stand-alone computer for $1,999.00, an optical mouse for $49.99, a plotter for $1,050.00, a laser printer for $299.99, and CAD software for $750.00.

   a. Calculate the total cost of this CAD station.   a. _____

   b. What is the total cost, excluding the cost of a printer?   b. _____

   c. What would it cost to equip five CAD drafters with a complete CAD station such as this?   c. _____

CAD Problems

14. All sections of the CAD drawing of the arch of this bridge are equal in length. Calculate dimensions **A** and **B**.

   A _____

   B _____

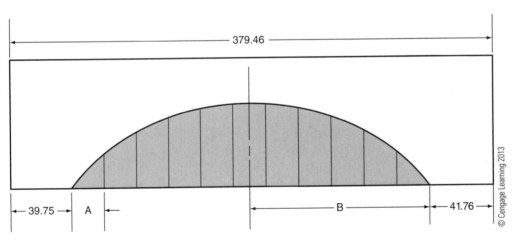

15. Calculate dimensions **A** and **B** on this CAD drawing of a template.
(NOTE: The symbol TYP in the drawing indicates that the same
dimension is found at the other end.)

A _____

B _____

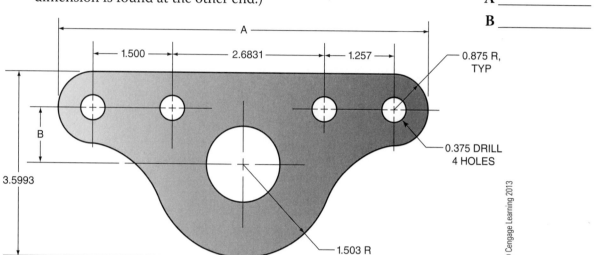

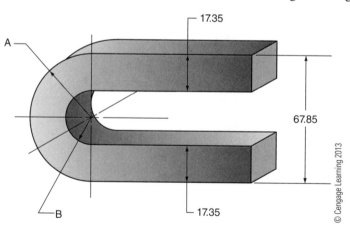

16. Calculate dimensions **A** and **B** on this CAD drawing of a magnet.

A _____

B _____

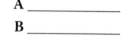

© Cengage Learning 2013

17. Calculate the overall lengths of this CAD drawing of a match plate and
    dimension **A.**   Match plate _____

                                                    Dimension A _____

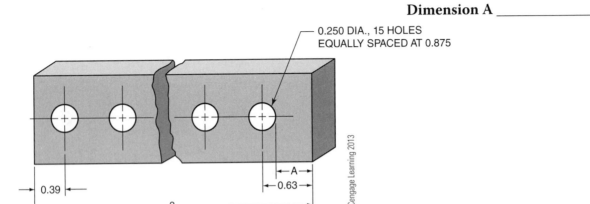

0.250 DIA., 15 HOLES
EQUALLY SPACED AT 0.875

0.39

?

A

0.63

© Cengage Learning 2013

18. The outside diameter (OD) of a section of pipe is 2.250″. The wall
    thickness is 0.387″. Calculate, in inches, the inside diameter (ID) of the
    pipe.   _____

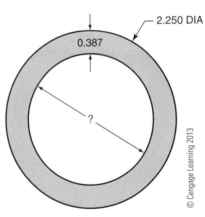

2.250 DIA

0.387

?

© Cengage Learning 2013

19. Three principal views of an object are centered on a 12 × 18 sheet of vellum. Calculate dimension **A** and dimension **B**.

A _____

B _____

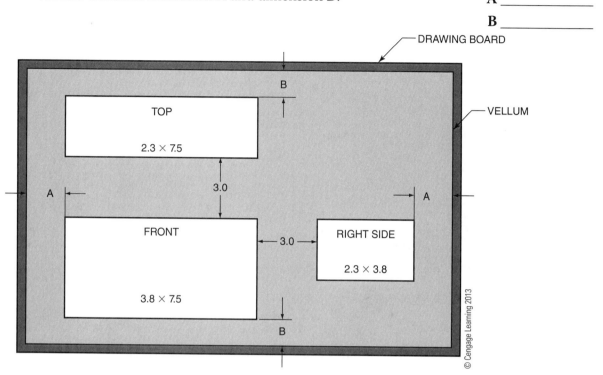

20. Use the CAD drawings below and calculate dimensions **A** and **B**. Express the answers rounded to the nearest thousandth.

A _____

B _____

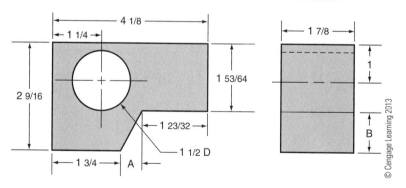

21. Using the CAD drawing and the dimensions below, calculate the overall height and width of this object.

Height _____

Width _____

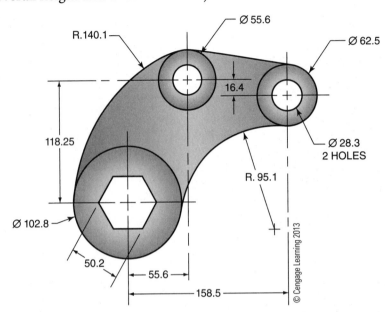

Ø 55.6

R.140.1

Ø 62.5

16.4

118.25

Ø 28.3
2 HOLES

R. 95.1

Ø 102.8

50.2

55.6

158.5

© Cengage Learning 2013

22. The height of the CAD drawing below is to be reduced by a factor of 3 and the width by a factor of 5 when the object is inserted into another drawing. What will be the new dimensions of its height and width?

Height _____

Width _____

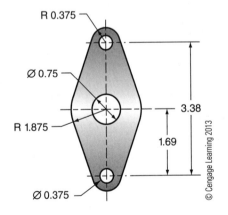

R 0.375

Ø 0.75

R 1.875

3.38

1.69

Ø 0.375

© Cengage Learning 2013

23. Use the CAD drawing below and calculate the new overall height and width if the drawing is enlarged by a scale factor of 1.75.

Height _____

Width _____

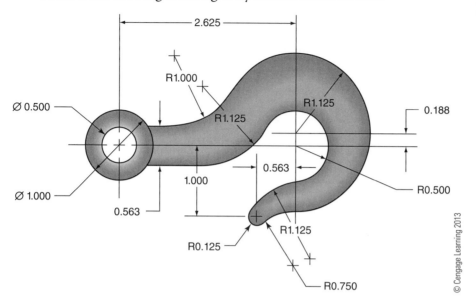

24. Use the CAD drawings of a locating block below to determine dimensions **A, B,** and **C.** Then determine the new height, width, and depth of the object if it is scaled down by a factor of 8.

A _____

B _____

C _____

New height _____

New width _____

New depth _____

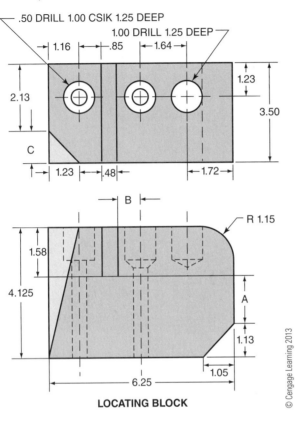

LOCATING BLOCK

25. Calculate the overall height and width of the front view of the CAD drawing below.

Height _____

Width _____

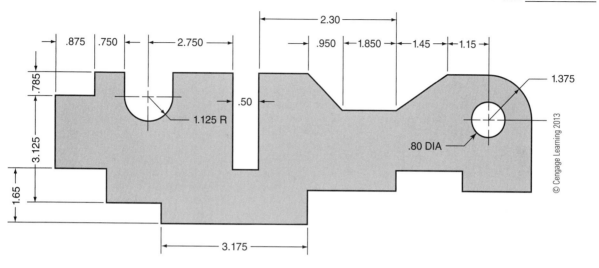

FRONT VIEW

26. Calculate dimensions **A, B, C,** and **D** using the CAD drawing below. Express the answers as decimals rounded to three decimal places.

A _____

B _____

C _____

D _____

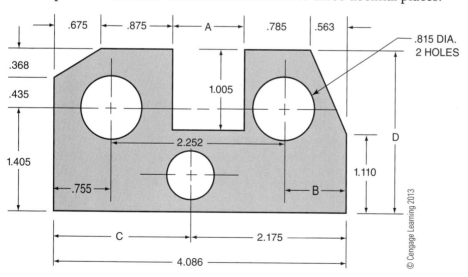

SECTION

SECTION 4

Decimals, Fractions, and Percents

UNIT 16

Equivalent Decimals and Fractions

Basic Principles

Equivalent numbers are numbers in any form that have the same value. This includes decimals and fractions such as $0.5 = \frac{1}{2}$ and $1.0625 = 1\frac{1}{16}$.

- A fraction can be expressed as a unique decimal that terminates (ends), such as $\frac{3}{4} = 0.75$, repeats endlessly, such as $\frac{1}{3} = 0.\overline{3}$ and $\frac{1}{7} = 0.\overline{143857}$, or continues on with no repeating pattern or end, such as 3.14159... and 1.4142..... A bar over one or more decimal digits in a number means that those digits continue to repeat. An ellipsis (...) following the last given place of a decimal means that the digits go on and on without repeating.

- A decimal can be expressed as one or more equivalent fractions or mixed numbers, such as $0.75 = \frac{75}{100} = \frac{15}{20} = \frac{3}{4}$ and $2.125 = 2\frac{1}{8} = \frac{17}{8}$. Remember: a fraction in simplest form, such as $\frac{3}{4}$, has many multiples—$\frac{6}{8}$, $\frac{15}{20}$, $\frac{30}{40}$, and so on.

Converting Fractions and Mixed Numbers to Decimals

To convert a fraction into its decimal equivalent, just divide the numerator by the denominator.

EXAMPLE: Convert the fraction ⅝ to its equivalent decimal.

$$
\begin{array}{r}
0.625 \\
8\overline{)5.000} \\
-4.8 \\
\hline
20 \\
-16 \\
\hline
40 \\
-40 \\
\hline
0
\end{array}
$$

So, $\dfrac{5}{8}$ = 0.625, which is a terminating decimal.

To convert a mixed number to its equivalent decimal, first write the mixed number as an improper fraction and then divide the numerator by the denominator.

EXAMPLE: Convert 4⅜ to its decimal equivalent.

STEP 1:   Write the mixed number as an improper fraction. $4\dfrac{3}{8} = \dfrac{35}{8}$

STEP 2:   Divide the numerator by the denominator. Carry out the division until the decimal digits end or repeat.

$$
\begin{array}{r}
4.375 \\
8\overline{)35.000} \\
-32 \\
\hline
30 \\
-24 \\
\hline
60 \\
-56 \\
\hline
40 \\
-40 \\
\hline
0
\end{array}
$$

So, $4\dfrac{3}{8}$ = 4.375, a terminating decimal.

Converting Decimals to Fractions

A decimal is a fraction that has a denominator that is a power of 10, such as 10, 100, 1000, and so on. The number of places to the right of the decimal point indicates the power of 10 in the denominator. The digits to the right of the decimal point form the numerator of the fraction.

- If the decimal has one or more zeroes immediately to the right of the decimal point followed by no non-zero digits, the zeroes can be dropped: for example, 3.00 = 3.
- To identify the power of 10 in the denominator of a decimal written as a fraction, count the number of places in the decimal number. The denominator is that power of 10. For example, 3.752 has three decimal places after the decimal point. So, the denominator of the fraction is 1000 and 3.752 = 3 752/1,000.
- In some cases, the resulting fraction can be reduced (simplified) to lowest terms.

EXAMPLES: Convert these decimals into equivalent fractions or mixed numbers in simplest form.

$$0.27 = \frac{27}{100}$$

$$0.063 = \frac{63}{1000}$$

$$3.9 = 3\frac{9}{10}$$

$$12.125 = 12\frac{125}{1000} = 12\frac{1}{8}$$

NOTE: Read 12.125 as *twelve and one hundred twenty five thousandths.*

Most calculators have a second function key $\boxed{\text{F↔D}}$ that lets you convert most fractions (F) to decimals (D) and vice versa. When this key is used, the equivalent decimal or fraction appears automatically in the display of the calculator.

NOTE: You may not always be able to use these functions on every calculator, however, if the number of digits and symbols exceeds the capability of the calculator's display. For example, some calculators cannot convert $^{531}/_{1000}$ to its decimal equivalent 0.531 because $^{531}/_{1000}$ has a total of eight characters, including the division symbol for the fraction bar (÷).

You can solve all of the problems in this unit by applying mathematical operations directly (using your head), or you can use either a calculator or TABLE II, FRACTION AND DECIMAL EQUIVALENTS on page 393 of this book. The proper fractions listed in TABLE II are those that occur most often in drafting problems and have denominators of 4, 8, 16, 32, and 64. Each fraction corresponds to an equivalent decimal having three or more places. However, some decimal equivalents have been rounded, which means that using the table may result in your getting approximations of an exact answer, either in fraction or decimal form; for example, $\frac{5}{32} = .15625$, which may be rounded to .1563. Using operations to calculate a value or using a calculator should give the exact answer. Therefore, for particular problems in this unit, you may find in the Answer Key and Instructor's Guide both the exact answer and an approximation based on the values in Table II. Either number is acceptable.

Skill Problems

Express the following fractions as equivalent decimals rounded to the nearest thousandth as necessary.

1. $\dfrac{2}{7}$ \underline{\hspace{4cm}}

2. $\dfrac{5}{16}$ \underline{\hspace{4cm}}

3. $\dfrac{11}{24}$ \underline{\hspace{4cm}}

4. $\dfrac{19}{32}$ \underline{\hspace{4cm}}

Express the following decimals as equivalent fractions in simplest form.

5. 0.1875 \underline{\hspace{4cm}}

6. 0.71875 \underline{\hspace{4cm}}

7. 0.8125 \underline{\hspace{4cm}}

8. 0.015625 \underline{\hspace{4cm}}

Practical Problems

9. The width of an erasing shield is $2\frac{5}{16}$". What is this width in decimal form rounded to the nearest thousandth inch? \underline{\hspace{4cm}}

10. An eraser is 0.475 inches thick. What is the thickness, in inches, written as a fraction in simplest form? _____

11. A civil drafter's dust brush weighs 1.25 lb. Calculate, in pounds, the weight of 23 brushes. Express the answer as a mixed number in simplest form. _____

12. A machinist uses a lathe to turn down a $^{31}\!/_{64}$" diameter shaft from a $^9\!/_{16}$" diameter piece of stock.

    a. Calculate, in inches, the fractional difference between the diameters. **a.** _____

    b. Express the fraction as a decimal rounded to the nearest thousandth. **b.** _____

13. A line is divided into three segments of $3^{49}\!/_{64}$", $2^{13}\!/_{32}$", and $1^{19}\!/_{64}$".

    a. What is the total length of the line in inches expressed as a mixed number in simplest form? **a.** _____

    b. Convert the sum to a decimal rounded to the nearest thousandth. **b.** _____

14. A visible object line on a drawing is approximately 0.034" thick. A hidden object line is approximately 0.018" thick. Calculate, in inches, the difference between the two thicknesses and express the answer as a fraction in simplest form. _____

CAD Problems

15. Calculate the decimal value of the perimeter of this CAD drawing of a die insert scale. Then express this value as a mixed number in simplest form. _____

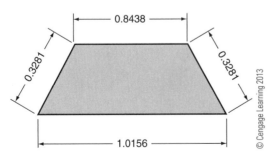

© Cengage Learning 2013

16. A regular pentagon has five equal sides. Calculate the perimeter of this
regular pentagon expressed as a mixed number in simplest form.   _____

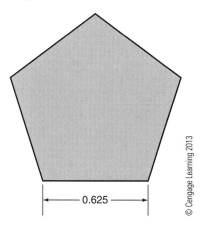

0.625

© Cengage Learning 2013

17. Calculate dimensions **A** and **B** on this CAD drawing expressed as
fractions in simplest form.

A _____

B _____

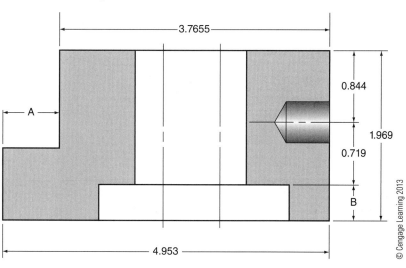

3.7655

0.844

1.969

0.719

A

B

4.953

© Cengage Learning 2013

18. Express the dimensions on this CAD drawing of a locating finger as
    decimals rounded to the nearest thousandths as necessary.

A _____

B _____

C _____

D _____

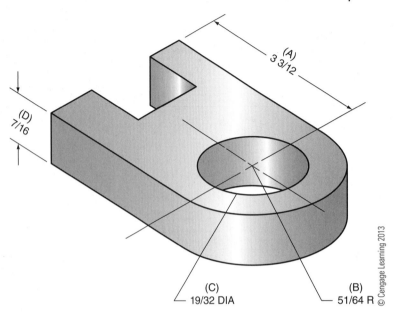

19. Before a tapping operation, a 0.53125″ diameter hole is drilled in
    the CAD drawing of the block below. The block is tapped with a ⅝″
    diameter tap drill. Calculate the difference in inches between the two
    diameters, expressed as a fraction in simplest form.

_____

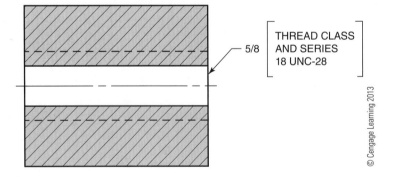

20. Calculate dimension **A** on the CAD drawings of a stepped shaft below, expressed as a fraction in simplest form.

_____

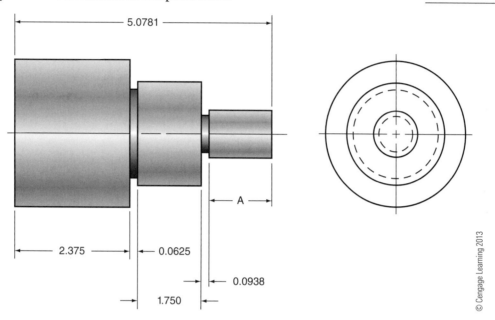

© Cengage Learning 2013

21. Use the CAD drawing below to calculate dimension **A**. Express the answer as a fraction in simplest form.

_____

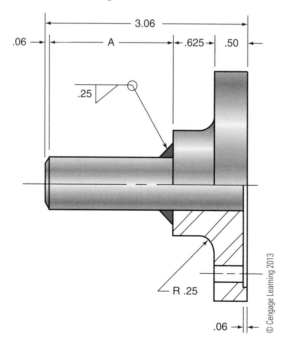

© Cengage Learning 2013

22. The 2-inch circles shown in the CAD drawing below are to be reduced by a factor of 6. Determine the new value in inches of the circles and express the answer as a fraction in simplest form.

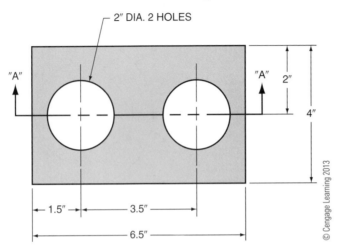

2″ DIA. 2 HOLES

"A"   "A"   2″

4″

1.5″   3.5″

6.5″

© Cengage Learning 2013

23. Use the CAD drawing below to calculate dimensions **A** and **F** if **B** is ⅝″, **C** is 1⅝″, **D** is 2²⁹⁄₃₂″, and **E** is 3⁵⁄₆₄″. Express **A** and **F** rounded to three decimal places.

A _____

F _____

A   G

W X   Y

B   C   E

D

F

© Cengage Learning 2013

24. Use the CAD drawing below to determine how far, in inches, from the center of the large circle the four small holes would be if they were moved out halfway between the two circular center lines (bolt circles). Express the answer as a mixed number in simplest form. _____

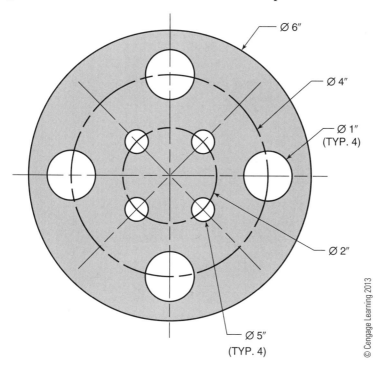

Ø 6"

Ø 4"

Ø 1"
(TYP. 4)

Ø 2"

Ø 5"
(TYP. 4)

© Cengage Learning 2013

25. Calculate dimensions **A** and **B** using the CAD drawing below. Express the answers as decimals rounded to the nearest thousandth as necessary.

A _____

B _____

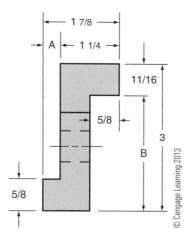

1 7/8

A ⊢ 1 1/4

11/16

5/8

3

B

5/8

© Cengage Learning 2013

26. Determine the overall length and height of the symmetrical CAD drawing below. State the answers as mixed numbers in simplest form.

Length _____

Height _____

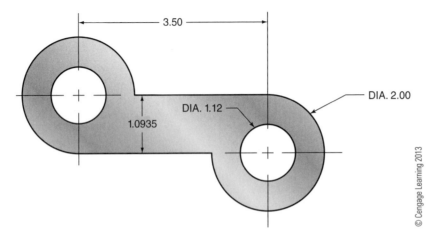

27. Use the CAD drawing below to determine dimension **A** expressed as a fraction in simplest form.

_____

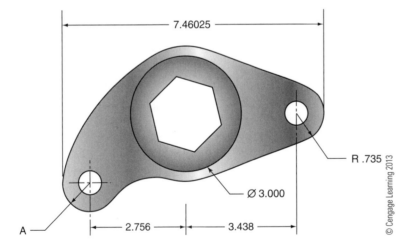

UNIT 17

Percents

Basic Principles

A *percent* is a fraction whose denominator is 100. The word *percent* and its symbol % mean *parts per hundred*.

$$\% = \frac{\text{part}}{100}$$

A number written as a percent, such as 25%, is equivalent to the fraction $\frac{25}{100}$, where 25 is the part, and 100 is the whole.

Because a percent represents a fraction in hundredths, it can also be written as a decimal.

EXAMPLE: $43\% = \dfrac{43}{100} = 0.43$

        🖩   43 ÷ 100 =

Converting Percents to Decimals

An easy way to convert a percent to a decimal is to drop the percent sign and move the decimal point in the number two places to the left. You may have to insert zeroes as shown in the first example below.

EXAMPLES: $5\% = 0.05$ $28.5\% = 0.285$ $100\% = 1$ $125\% = 1.25$

Converting Decimals to Percents

To convert a decimal number to a percent, move the decimal point in the number two places to the right and add the percent sign. NOTE: You may have to add a zero as in the second example below.

EXAMPLES: $0.015 = 1.5\%$ $0.8 = 80\%$ $0.35 = 35\%$ $2.125 = 212.5\%$

Converting Fractions to Percents

To convert a fraction to a percent, convert the number to its decimal equivalent by dividing and then write the decimal as a percent as shown above. You may have to add zeroes as in the third example below.

EXAMPLES: $\dfrac{3}{4} = 0.75 = 75\%$ $\dfrac{5}{8} = 0.625 = 62.5\%$ $3\dfrac{1}{2} = 3.5 = 350\%$

Converting Percents to Fractions

To change a percent to a fraction, write the percent as a fraction whose denominator is 100 and simplify the fraction to simplest form as necessary.

EXAMPLE: $40\% = \dfrac{40}{100} = \dfrac{4}{10} = \dfrac{2}{5}$

The Three Cases of Percents

The definition of *percent* relates three quantities: a percent, a part, and a whole. If you know any two of these quantities, you can solve for the third one. Here are the rules for dealing with each of the three types (cases) of percent problems.

CASE 1:   If you know the values of the part and the whole, set up the fraction $\frac{part}{whole}$ and then divide. Multiply the quotient by 100 to calculate the percent.

EXAMPLE: What percent of 60 is 15 feet?

The part is 15, and the whole is 60. Calculate the percent.

$$\text{Percent} = \dfrac{15}{60} \times 100$$

$$= \dfrac{1}{4} \times 100$$

$$= 25$$

So, 15 ft is 25% of 60 ft.

You can check the answer by multiplying 60 and 25%: $60 \times 0.25 = 15.$ ✓

CASE 2: If you know the percent and the whole, write the percent as a decimal (or fraction) and multiply the whole by the decimal (or fraction) to calculate the part.

EXAMPLE: What is 25% of 24 inches?

The percent is 25, and the whole is 24. Calculate the part.

$$25\% = \frac{\text{part}}{24}, \text{ where } 25\% = .25 = \frac{1}{4}.$$

$$\text{Part} = 0.25 \times 24 = \frac{1}{4} \times 24 = 6$$

So, 25% of 24 in. is 6 in.

You can check the answer by multiplying 25% and 24: $0.25 \times 24 = 6$. ✓

CASE 3: If you know the percent and the part, you can find the whole by dividing the part by the percent written as a decimal.

EXAMPLE: 48 is 75% of what number?

The percent is 75, and the part is 48. Calculate the whole.

$$75\% = \frac{48}{\text{whole}}, \text{ where } 75\% = 0.75$$

$$\text{whole} = \frac{48}{0.75} = 64$$

So, 75% of 64 is 48.

You can check the answer by multiplying 75% and 64: $0.75 \times 64 = 48$. ✓

NOTE: When working with percent problems, you can think of the word *of* as meaning *times,* which indicates multiplication.

EXAMPLE: 15% *of* 60 → 15% × 60

Skill Problems

Calculate the following percentages. Express answers to the nearest hundredth where necessary.

1. 7% of 48   _____

2. 12% of 86   _____

3. 20% of 75   _____

4. 35% of 197   _____

5. 21% of 1192   _____

6. 87¼% of 724   _____

7. Shade 75% of the figure below. (Hint: The whole is the total number of unit squares inside the figure.)   _____

© Cengage Learning 2013

8. Shade 37.5% of the figure below.   _____

© Cengage Learning 2013

Practical Problems

9. How many sections of the circle below should be shaded to represent 75% of the circle?

_____

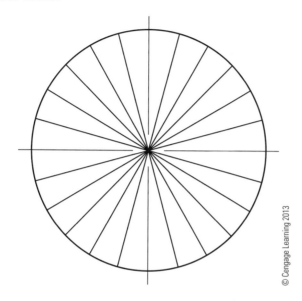

© Cengage Learning 2013

10. A checker in a structural drafting department checked 150 drawings in one month. Twenty-two percent of the drawings had more than 12 mistakes. How many drawings had more than 12 mistakes?

_____

11. A CAD drafter predicts that a specific job should take 109 hours. The number of hours he works to complete the job is 7% less than the predicted time. Calculate the number of hours the CAD drafter takes to complete this job.

_____

12. A machined block that is 3.345″ thick must be reduced by 25%. What is the new thickness in inches after the reduction? Round the answer to the nearest hundredth.

_____

13. An architectural drafting department supervisor orders 136 reams of assorted sizes of paper. When the order is delivered, 12½% of the total order is on back order. How many reams of paper were not delivered?

_____

14. A CAD drafter invests 35% of his $25,000 savings in a real estate property. Determine the amount of money he invested in this property.

_____

15. A large drafting department consists of 368 drafters. During a 2-year period, 6¼% of the drafters retired. How many of the original drafters remain in the department?

_____

16. A drafting job was completed with standard drafting equipment in 60 hours. A drafting machine with a protractor head is said to decrease a drafter's time by 30%. Based on this percentage, how many hours would be saved if a drafter used a drafting machine?

_____

17. Of the 150 drafters employed by a company, 18% have part-time jobs. How many drafters have part-time jobs?

_____

18. A large engineering firm employs 76 drafters. There are 12 structural drafters, 19 mechanical drafters, 14 civil drafters, and 25 architectural drafters. The rest of the employers are CAD drafters. Calculate the approximate percentage for each classification of drafter, rounded to the nearest whole number.

% Mechanical _____   % Civil _____

% Architectural _____   % CAD _____

% Structural _____

19. A CAD drafter assigns 200 layers of the 256 available layers in his CAD program in the following manner: 75 for plan views, 50 for elevations, 40 for details, 25 for construction, and 10 for miscellaneous specifications. Calculate, to the nearest whole number, the percentage of each layer to be used and the percentage of layers assigned from the total number of layers.

% Total layers _____   % Plan views _____

% Elevations _____   % Details _____

% Construction _____   % Miscellaneous _____

20. A manufacturing company has 36 drafters, 75% of whom use CAD rather than more traditional ways to produce their drawings. How many drafters do not use CAD in this company?

_____

21. A CAD department supervisor earns a weekly gross income of $1200.00. He wants to calculate the amount of tax that is deducted from his paycheck for Social Security (FICA) and Medicare. The combined tax deduction for these two programs is 7.65% of his gross income. Determine the amount of this monthly deduction.

_____

22. A CAD drafter wants to determine her gross income, taxable income, and withholding tax for the previous week. The drafter worked 32 hours last week at a rate of $22.50 per hour. Her withholding income is 28.0% of her gross pay less any deductions. She has one dependent for which the IRS (Internal Revenue Service) allows a deduction of $50 from her taxable income per week. Determine her gross income, her taxable income, and her withholding tax for last week.

Gross income _____

Taxable income _____

Withholding tax _____

23. A drafter earns a gross income of $1300 per week and has three dependents. She wishes to verify her net income per week after all deductions are taken.

a. A person's weekly *taxable income* is the gross income less weekly deductions of $50 for each dependent. What is the drafter's taxable income?   a. _____

b. The deduction for her withholding tax is 25% of her weekly taxable income. What is her withholding tax?   b. _____

c. The Social Security and Medicare deduction is 7.65% of her weekly gross income. What is this total deduction?   c. _____

d. What is the drafter's net income per week?   d. _____

24. The total cost of a drafting subcontractor's job was $426.00. The materials cost 6% of the total amount, the computer time costs 14% of the total amount, and the remaining costs were charged to labor. What are the amounts for each of the three costs for this job?

Materials _____

Computer time _____

Labor _____

CAD Problems

25. A designer must redesign the CAD drawing of the plate below so that dimension **A** will be 5% smaller, and dimension **B** will be 7% larger. What are the new dimensions of **A** and **B**? Round the answers to the nearest thousandth.

A _____

B _____

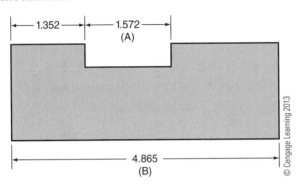

26. Because of an incorrect dimension on a detail drawing, a drafter must reduce the hole diameter in this CAD drawing of a stop block by 15%. What is the diameter of the new hole, rounded to the nearest thousandth?

_____

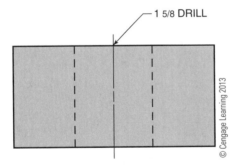

27. The overall length of the CAD drawing of the T-guide below must be enlarged by 7¼%. What is the new length of the guide rounded to the nearest thousandth?

_____

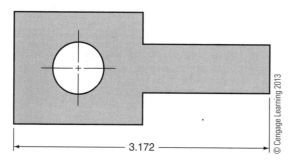

3.172

© Cengage Learning 2013

28. A CAD drafter needs to enlarge the CAD drawing below so it will fit on top of another part. The height will be enlarged by 15% and the width by 20%. Determine the dimensions of the new height and width, rounding the answers to the nearest hundredth as necessary.

Height _____

Width _____

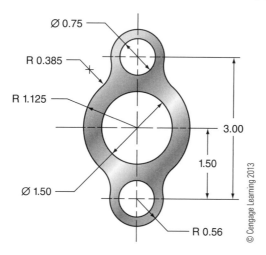

Ø 0.75

R 0.385

R 1.125

Ø 1.50

R 0.56

3.00

1.50

© Cengage Learning 2013

UNIT 18

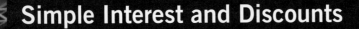

Simple Interest and Discounts

Basic Principles

Calculating simple interest and discounts are ways of using percentages. In the case of interest, you pay an extra amount of money based on a purchase or a loan. In the case of a discount, you pay less for a purchase.

Simple Interest

When you borrow money, the amount borrowed (the loan) is called the *principal*. The *simple interest* is a *rate* expressed as a percent over time.

The amount of simple interest charged for a loan is the product of three quantities: the principal, the rate of interest per unit time, and the time period of the loan. When the term of a loan has expired, the money you repay equals the principal plus the interest accrued or owed.

Formulas

1. Simple interest (I) = principal (p) × rate (r) × time (t)
   $I = prt$
2. Amount to be repaid (A) = principal (p) + interest (I)
   $A = p + I$

EXAMPLE: A company borrows $15,000.00 at a simple interest rate of 12% per year over a 2-year period. How much money is repaid after two years?

STEP 1:   Calculate the amount of simple interest on the loan. (12% = 0.12)

$I = prt$ $I = \$15,000 \times 12\%/\text{yr}. \times 2$ yr.

$p = \$15,000$ $I = \$15,000 \times 0.12 \times 2$

$r = 12\%/\text{yr}.$ $I = \$3,600$

$t = 2$ yr.

STEP 2:   Calculate the amount to repay: the principal (p) plus interest (i).

$\$15,000 + \$3,600 = \$18,600$

The amount to be repaid for this loan is $18,600.00.

▦   15,000 ⊠ .12 ⊠ 2 ⊞ 15,000 ▱

NOTE: If your calculator has a % key, you can also enter the following expression.

▦   15,000 ⊠ 12 2nd % ⊠ 2 ⊞ 15,000 ▱

Discounts

A discount is an amount subtracted from the cost of something. A discount is usually expressed as a percentage of a *list price.* To calculate a discount, multiply the discount rate and the list price. The *net price* is the list price minus the discount.

Discount Formulas

1. Discount (D) = Rate (r) × Cost (C)
    $D = r \times C$
2. Net price (N) = Cost (C) − Discount (D)
    $N = C - D$

EXAMPLE: A cost of a set of drafting instruments lists for $38.00. The original selling price has been discounted by 10%. An additional 3% discount is offered if payment is in cash. Calculate the net cost if a drafter pays for the set of instruments in cash.

METHOD 1

STEP 1:   Calculate the 10% discount on the list price. (10% = 0.10)

$D = 0.10 \times \$38.00 = \$3.80$

STEP 2:   Subtract to calculate the net price after the first discount.

$\$38.00 - \$3.80 = \$34.20$

STEP 3:   Multiply the net price by 3%. (3% = 0.03)

$D = 0.03 \times \$34.20 = \$1.03$

STEP 4:   Subtract the second discount from $34.20 to determine the net cost of the purchase after both discounts.

$\$34.20 - \$1.03 = \$33.17$

The net cost is $33.17.

METHOD 2

A discount of 10% means that the drafter pays 90% of the selling price (100% − 10% = 90%).

STEP 1:   Multiply the list price by 90%. (90% = 0.90)

$0.90 \times \$38.00 = \$34.20$

STEP 2:   A discount of 3% means that the drafter pays just 97% of the sales price, $34.20. So subtract 3% from 100%.

$100\% - 3\% = 97\%$

STEP 3:   Multiply the new sale price, $34.20, by 97%. (97% = 0.97)

$0.97 \times \$34.20 = \$33.17$

The net cost is $33.17.

Practical Problems

1. The list price of a computer is $4,800.00. Calculate the cost of the computer if the vendor is advertising a discount of 10%.

   _____

2. The list price of a computer program is $178.00 with a 15% discount, followed by a discount of 2% if paid in cash. Calculate the net cost of a cash purchase.

   _____

3. A civil engineer purchases an item through a wholesale outlet, paying 87% of its retail price. What was the percent discount for the item?

   _____

4. The steel bar below weighs 2.5 pounds per foot and costs $1.48 per pound. The given dimension is 30 inches. Calculate the costs of four lengths of steel before and after a discount of 9% is applied.

   Pre-discount
   cost _____

   Discounted
   cost _____

5. A package of 250 sheets of drafting vellum costs $17.50.

   a. Determine the cost of each sheet of vellum.

   a. _____

   b. What is the cost of 12 packages of vellum after a discount of 7% is applied?

   b. _____

6. The cost of a drafting instrument set is $21.50. A 5½% discount is given if a buyer pays cash. Determine the net price, to the nearest cent, if a buyer pays cash for the instrument set.

   _____

7. The list price for T-squares is $38.95 per dozen. There is a trade discount of 19% per dozen. If a company buys a dozen T-squares, calculate the cost, to the nearest cent, of each T-square.

   _____

8. A 23% discount is to be made on a purchase that costs $196.78. Calculate the discounted price to the nearest cent.

   _____

9. A manufacturer calculates the following costs for a drafting project: $27.50 for materials, $425.00 for labor, and $135.00 for overhead. If the profit for the project is 37% of the total cost for the project, calculate to the nearest cent what the manufacturer must charge the customer to get this profit. _____

10. The CAD drawing of a special nozzle fitting below sells for $42.72 per dozen, less three discounts of 10%, 6%, and 2%, applied one after the other.

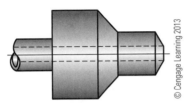

© Cengage Learning 2013

    a. Determine the final net price, to the nearest cent, of a dozen fittings.   a. _____

    b. Determine the net price, to the nearest cent, of each fitting after the three discounts have been applied.   b. _____

11. The total cost of a drafting package includes three categories: materials cost ($12.68), a labor cost ($1,255.00), and an overhead cost ($375.00). Calculate, to the nearest whole number, the percent of the total cost of the package for each category.

                                                          % Material cost _____

                                                          % Labor cost _____

                                                          % Overhead cost _____

12. Calculate the final net cost of 3600 pounds of metal at 38¢ per pound, less three discounts of 12%, 9%, and 3% applied one after the other. _____

13. The cost of manufacturing 1000 shims such as the CAD drawing of the one below is $863.00. Of this cost, 31% is spent for materials, 48% for labor, and the remainder for overhead.

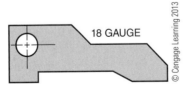

18 GAUGE

© Cengage Learning 2013

    a. What is the materials cost?   a. _____

    b. What is the labor cost?   b. _____

    c. What is the overhead cost?   c. _____

14. The list price of a complete set of drafting instruments is $32.45. The set is sold with a 14% discount. What is the net cost of the instrument set? _____

15. A company purchased 24 set-up blocks like the CAD drawing of the one below. A set of four set-up blocks sells for $85.60. At the time of purchase, a 12% trade discount, followed by a 4% cash discount, was allowed. Calculate each of the following expenses:

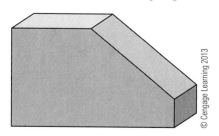

© Cengage Learning 2013

a. the cost of 24 set-up blocks before the discounts were applied.   a. _____

b. the cost of 24 set-up blocks if a company pays in cash.   b. _____

c. the net cost to the company for each set-up block purchased.   c. _____

16. The monthly payroll for a small drafting department is $24,500. The monthly payroll for the company is $68,571.00. Calculate the percent of the company's monthly payroll that the drafting department payroll represents. Express the answer to the nearest whole number percent. _____

17. The average yearly operating cost of a drafting department is $15,200.00 per person. Calculate the average operating cost for the year if the drafting department employs 12 workers. _____

18. The cost of manufacturing the CAD drawing of the step shaft below is reduced 17% by eliminating the undercuts at the shoulders. The original cost of the shaft is $9.37.

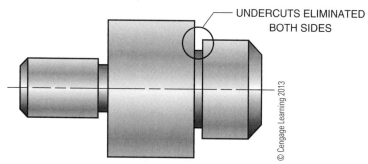

UNDERCUTS ELIMINATED
BOTH SIDES

© Cengage Learning 2013

a. Calculate the savings per individual part.   a. _____

b. Calculate the net cost of producing the step shaft.   b. _____

19. How much interest will $4,500.00 earn in one year if the simple interest rate is 3½%? _____

20. A loan of $54,000.00 is taken out to enlarge a drafting department. The simple interest rate on the loan is 12% per annum. If the debt is repaid in 18 months, how much simple interest was charged on the loan? _____

21. A loan is secured for $7,600.00 at a rate of 8% per annum. The loan is paid when it becomes due at the end of 27 months. Calculate the total amount of simple interest charged on the loan. _____

22. The annual rate of simple interest on a loan of $1,500 is 17%.

    a. What is the interest after a period of three months?   a. _____

    b. After a period of four months?   b. _____

23. Calculate the total amount repaid on a loan of $8,600.00 at a rate of 15% simple interest after three years. _____

24. A bank loans a drafter $9,000.00 for which $675.00 interest is to be paid annually. What is the rate of simple interest on the loan? _____

25. A family pays $5,490.00 simple interest annually on a mortgage of $36,000.00. What is the rate of simple interest? _____

26. Calculate the simple interest on $3,735.75 for one year at a rate of 9½%. _____

27. A bank note is issued for $3,725.00 at a simple interest rate of 3% per month. How much interest is charged for five months? _____

28. An engineering firm adds an annual 18% interest charge (1.5% per month) on money owed them after 30 days. How much simple interest is owed on a bill of $3,750.00 that is paid in 90 days? _____

29. A CAD drafter deposits 12% of his yearly income in a special bank account that earns interest at a rate of 6.5% per annum. His yearly salary is $65,000.00. How much simple interest will this investment earn in two years? _____

30. A stand-alone CAD station sells for $4,575.00. A CAD drafter who purchases the station receives a discount of 7.5% off the sales price. What does the drafter pay for the CAD station after the discount is applied? _____

Geometry

UNIT 19

Powers, Roots, and the Pythagorean Theorem

Basic Principles

Powers and roots of numbers are often used to calculate distances between different parts of machine drawings. Raising a number to a power involves multiplication. Finding the root of a number involves finding equal factors of a number. These two operations are *opposites*.

Powers of Numbers

The power of a number is represented by the *exponent* of the number. The number being raised to a power is called the *base.* When the exponent is a positive whole number, {1, 2, 3, …}, it tells you how many times a base is used as a factor.

$$5^3 = 5 \times 5 \times 5$$

exponent

base

Three factors of 5

Any whole number, fraction, or decimal can be raised to a power.

EXAMPLES:

$$2^6 = 2 \times 2 \times 2 \times 2 \times 2 \times 2 = 64$$

$$\left(\frac{3}{4}\right)^2 = \frac{3}{4} \times \frac{3}{4} = \frac{9}{16}$$

$$0.125^3 = 0.125 \times 0.125 \times 0.125$$
$$= 0.001953125$$

 2 y^x 6 $=$ (3 ÷ 4) x^2 2nd F↔D $=$ .125 y^x 3 $=$

NOTE: You can use the y^x key of a calculator to raise a number to a power. The variable *y* represents the base, and the variable *x* represents its exponent. To square a number, you can also use the x^2 as in the second example above.

190

Roots of Numbers

Finding the root of a number is the inverse (opposite) operation of finding the power of the number. The most common type of root is the *square root,* also called the *principal square root.* Its symbol is the *radical sign,* which looks like this: $\sqrt{}$. The number under a radical sign is called the *radicand.*

The principal square root of a number is always the positive number, which times itself, equals (or approximately equals) the radicand.

EXAMPLES: $\sqrt{4} = 2$ because $2 \times 2 = 4$. $\sqrt{0.25} = 0.5$ because $0.5 \times 0.5 = 0.25$.

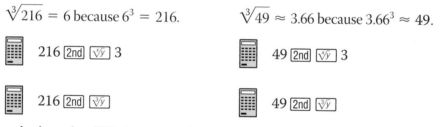

To indicate a root other than the square root of a number, a small number is written above the radical sign. This number is called the *index.* The index indicates what root is to be found. If there is no index number, it is assumed to be 2, which means find the square root of the number. If the index is 3, for example, the root of the radicand is the number that when raised to the third power (cubed) gives you the radicand.

$\sqrt[3]{216} = 6$ because $6^3 = 216$. $\sqrt[3]{49} \approx 3.66$ because $3.66^3 \approx 49$.

216 [2nd] [$\sqrt[x]{y}$] 3   49 [2nd] [$\sqrt[x]{y}$] 3

216 [2nd] [$\sqrt[3]{y}$]   49 [2nd] [$\sqrt[3]{y}$]

NOTE: In the calculator key [$\sqrt[x]{y}$] above, *y* is the radicand, and *x* is the index. Enter the radicand first, then the index. You may also use the [$\sqrt[3]{y}$] key to find the cube root of a number directly.

If you don't have a calculator, you can use TABLE III, POWERS AND ROOTS OF NUMBERS (1 TO 100) in Section II of the Appendix on page 394 to find some powers and roots of numbers. To find the power of a number, you can multiply. To find the root of a number, you can use *trial and error.*

To use trial and error, choose a number that you think might be a root and raise it to the power indicated by the index of the radical. Compare that result with the radicand and adjust the number until the result is either equal to, or very close to, the radicand. Because the roots of most numbers are not whole numbers, you may have to round the result to a designated decimal place.

EXAMPLE: Approximate $\sqrt[3]{502}$ to the nearest whole number.

STEP 1:   The index of the radical is 3. So, choose a number, say 15, that might be the cube root of 502 and raise it to the third power. Use TABLE III, a calculator, or multiply.

$15^3 = 3,375$, which is greater than 502.

STEP 2:   Choose a whole number about halfway between 1 and 15 to cube, say 7.

$7^3 = 343$, which is less than 502.

STEP 3:   Choose a whole number between 7 and 15 and cube it until you find a result that is very close to 502. Try 8.

$8^3 = 512$, which is very close to 502. So, it's a reasonable whole-number approximation of $\sqrt[3]{502}$.

So, $\sqrt[3]{502}$ is about equal to 8.

The Pythagorean Theorem

One of the most common applications of powers and roots in drafting is the Pythagorean theorem. You can use this theorem when working with right triangles. The theorem allows you to calculate the length of one of the three sides of a right triangle if you know the lengths of the other two sides.

The longest side of a right triangle is called the *hypotenuse* and in the equation below is represented as c. The two shorter sides are called *legs* and are represented by a and b in the equation.

The Pythagorean theorem: In right triangle ABC, $c^2 = a^2 + b^2$

In words, this theorem states that in a right triangle, the square of the hypotenuse equals the sum of the squares of the other two sides.

Calculating the Length of a Hypotenuse

EXAMPLE: Use the Pythagorean theorem to find the length of dimension C in the right triangle below. Round the answer to the nearest tenth.

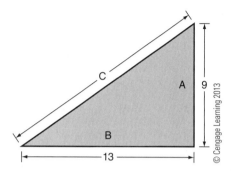

© Cengage Learning 2013

STEP 1: C is the hypotenuse, so substitute dimensions A and B into the equation $c^2 = a^2 + b^2$ for the legs a and b.

$$C^2 = 9^2 + 13^2$$

NOTE: The order of the sum of the squares of A and B is not important.

STEP 2:   Square each number first, then add the squares.

$$C^2 = 81 + 169 = 250$$

STEP 3:   To determine C, approximate the square root of 250 using a calculator, TABLE III, or trial and error. Then round the answer to the nearest tenth.

NOTE: There is no whole number squared that equals 250, so the answer is an approximation.

$$\text{So, } C = \sqrt{250} \approx 15.8.$$

Calculating the Length of a Leg

If the lengths of the hypotenuse and one leg are known, you can write the Pythagorean theorem in one of the following ways. Then solve for the length of either a or b by subtracting the squares first and then finding the square root of the difference.

$$a^2 = c^2 - b^2 \implies a = \sqrt{c^2 - b^2}$$
$$b^2 = c^2 - a^2 \implies b = \sqrt{c^2 - a^2}$$

EXAMPLE: In this right triangle, leg b is 7, and the hypotenuse c is 12. Determine the length of leg a. Round the answer to the nearest tenth.

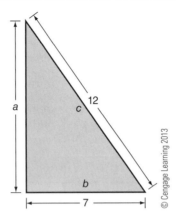

© Cengage Learning 2013

STEP 1:   To calculate a, apply the Pythagorean theorem, where c is the hypotenuse and b is the other leg.

$$a = \sqrt{c^2 - b^2}$$

STEP 2:   Substitute the known values of c and b into the equation, in that order.

$$a = \sqrt{12^2 - 7^2}$$

STEP 3:   Square 12 and 7 and subtract to simplify the radicand.

$$a = \sqrt{144 - 49} = \sqrt{95}$$

STEP 4:   Find the square root of 95 to the nearest tenth by using TABLE III, a calculator, or trial and error. Round the answer to the nearest tenth.

$$a = \sqrt{95} \approx 9.7$$

So, a is about equal to 9.7.

$12\boxed{x^2}\ \boxed{-}7\boxed{x^2}\ \boxed{=}\ \boxed{\sqrt{x}}$

Skill Problems

Raise each number to the indicated power.

1. 3^2 _____

2. 7^2 _____

3. 9^2 _____

4. 6^3 _____

5. 4^3 _____

6. 5^4 _____

7. 13^3 _____

8. 3.15^2 _____

9. 2.5^3 _____

10. 1.5^4 _____

11. 0.75^3 _____

12. $\left(\dfrac{3}{4}\right)^3$ _____

Find the indicated root of each number. Round answers to the nearest hundredth as necessary.

13. $\sqrt{49}$ _____

14. $\sqrt{169}$ _____

15. $\sqrt{6.819}$ _____

16. $\sqrt{875}$ _____

17. $\sqrt{1.898}$ _____

18. $\sqrt{746}$ _____

19. $\sqrt{\dfrac{0.81}{0.49}}$ _____

20. $\sqrt{76.43}$ _____

21. $\sqrt{217.5625}$ _____

22. $\sqrt[3]{343}$ _____

Practical Problems

23. The formula to calculate the area A of a square is $A = s^2$, where s is the length of each side. If the area of a square is 68.39 square inches, what is the length of each side of the square to the nearest hundredth inch? _____

24. An architectural drafter needs to find the length of the sides of a square-shaped room. The area of the floor is 135.75 square meters. Express to the nearest hundredth the length in meters of the sides of the floor. _____

25. A landscape architect plans a circular reflection pool for a park. The regulations will not permit a pool that has an area greater than 254.4696 square meters. Another formula to calculate the area A of a circle is $A = 0.7854 \times D^2$, where D represents the length of the diameter of the circle. Use this formula and determine the greatest length of a diameter for this reflecting pool.

_____

CAD Problems

26. What is the area in square units of the circle below, expressed to the nearest hundredth?

_____

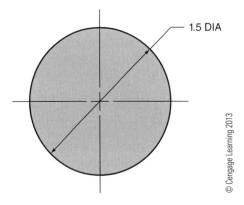

1.5 DIA

© Cengage Learning 2013

27. The volume V of a cube is equal to L^3, where L is the length of each edge of the cube. What is the volume of this cube expressed to the nearest tenth cubic unit?

_____

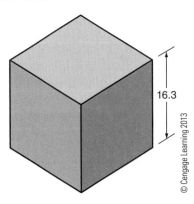

16.3

© Cengage Learning 2013

28. In this right triangle, side **BC** equals 10 inches, and side **AB** equals 15 inches. Calculate the length of side **AC** to the nearest hundredth inch. _____

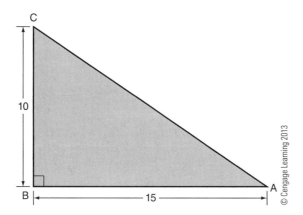

29. Use the given measures and calculate height **B** in inches on this CAD drawing of a gusset plate.   _____

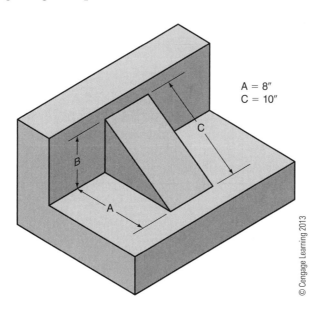

A = 8″
C = 10″

30. Calculate to the nearest hundredth, the thickness (dimension *A*) of this piece of steel. Use the Pythagorean theorem, written as $A = \sqrt{B^2 - C^2}$, where *A* is the height of the object, *B* is the diagonal of one of its faces, and *C* is the depth.

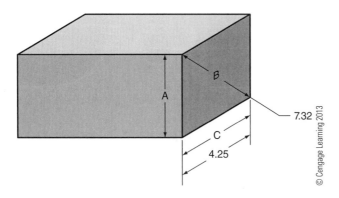

31. The CAD drawing below shows the left-profile view of a step-up block. What is the length of *B* rounded to the nearest hundredth? Use the Pythagorean theorem in the form $B = \sqrt{C^2 - A^2}$, where *A* and *B* are legs of the right triangle, and *C* is the hypotenuse.

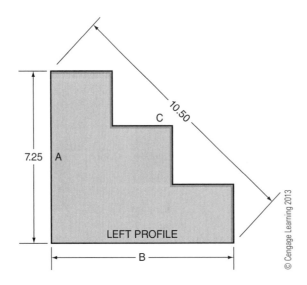

32. A mechanical drafter needs to calculate the length of dimension A in the CAD figure below. What is A?

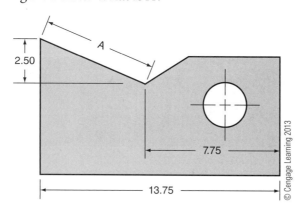

33. A CAD drafter needs to determine the lengths of two inclined lines AB and CD on the drawing below. Determine AB and CD to the nearest hundredth. (Hint: AB and CD are the hypotenuses of two right triangles.)

AB _____

CD _____

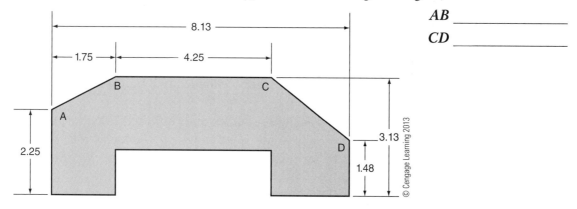

34. The CAD figure below is a trapezoid with two right angles. Calculate to the nearest hundredth, the value of X.

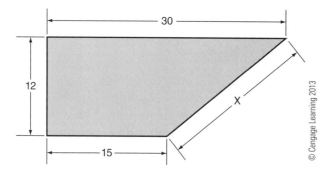

UNIT 20

Basic Principles

Almost all drafting figures and diagrams can be broken down into common geometric shapes, appearing as either two-dimensional or three-dimensional objects. To make correct drawings, a drafter must be able to recognize and construct different types of objects. This unit provides an overview of the most basic two-dimensional objects—lines, angles, and triangles—and their properties.

Lines

Plane geometry concerns figures in a *plane*, a flat surface made up of points and lines. Study the following properties of lines in a plane.

- A straight line has only one dimension—length—and extends infinitely in opposite directions.
- Two straight lines intersect in exactly one point called the *point of intersection.*
- Two or more lines that intersect to form right angles are called *perpendicular lines.* The symbol for perpendicularity is ⊥.
- Two or more lines in a plane that do not intersect are called *parallel lines.* The symbol for parallelism is //.

NOTE: In the drafting world, a *line* is always a *line segment*; a part of a line that includes its endpoints and all points between them. This means that a *line* in drafting and CAD design always has a measurable length.

Angles

Angles are formed when two lines intersect. The size of an angle depends on the opening between its sides; it does NOT depend on the lengths of its sides. An angle is measured in degrees (°), and the number of degrees is related to the number of degrees in a circle, 360°.

Angles in geometry are grouped into four categories depending on their measures: *right angles, acute angles, obtuse angles,* and *straight angles.* In the figures below, $\overrightarrow{OA}$ and $\overrightarrow{OB}$ are the sides of $\angle O$ (shaded). Point O, the center of a circle, is called the *vertex* of the angle.

A right angle measures 90°.	An acute angle has a measure less than 90°.	An obtuse angle has a measure greater than 90° and less than 180°.	A straight angle measures 180°.

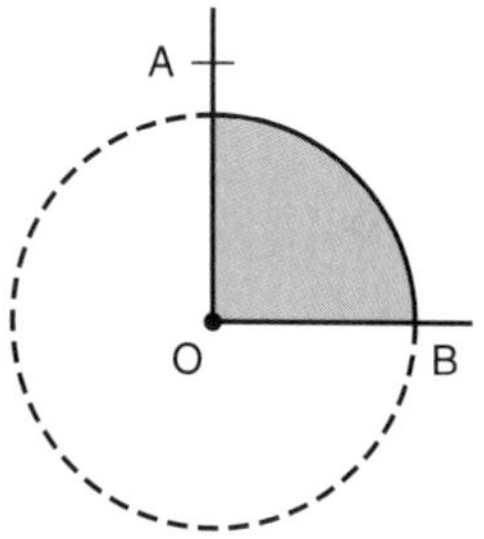

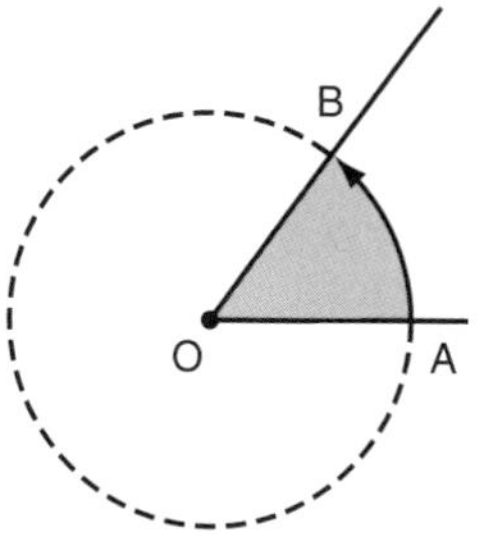

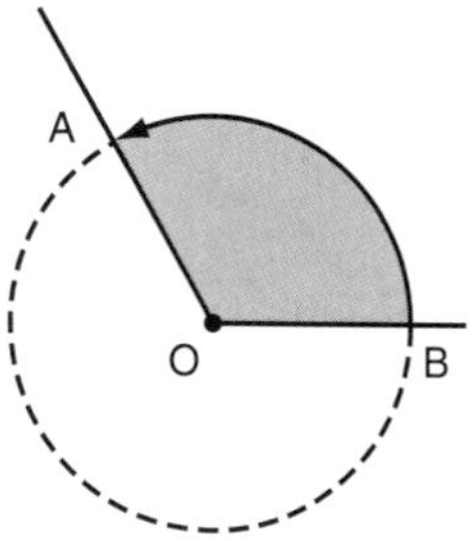

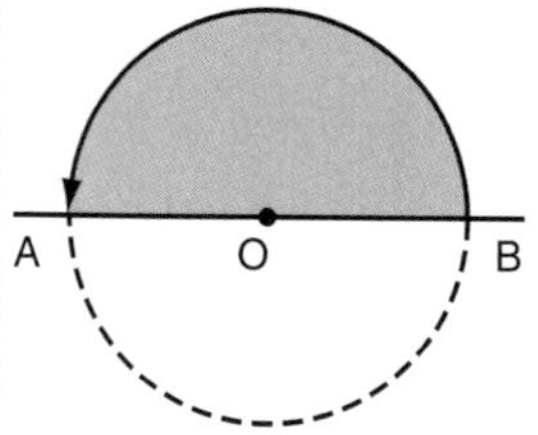

© Cengage Learning 2013

Triangles

Triangles are formed by three line segments that intersect at their endpoints. Every triangle has three sides and three angles and can be classified in terms of its sides, angles, or both; for example, an isosceles right triangle.

Name by sides	Property
Scalene	No two sides equal
Isosceles	Two sides equal
Equilateral	Three sides equal

Name by angles	Property
Acute	All angles acute
Right	One right angle
Obtuse	One obtuse angle

© Cengage Learning 2013

The following properties apply to all triangles.
- The sum of the measures of the angles in a triangle is 180°.
- The *perimeter* of a triangle is the sum of the lengths of its three sides.
- The sum of the lengths of two sides of a triangle is always greater than the length of the third side.
- The largest angle in a triangle lies opposite the longest side; the longest side lies opposite the largest angle.
- If the measures of three sides and three angles of one triangle are equal to the measures of three sides and three angles of another triangle, the triangles are said to be *congruent.*

Skill Problems

1. What type of angle is ∠*AOB*?   _____

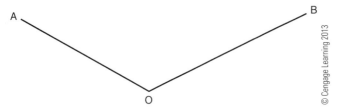

2. What type of angle is ∠*COD*?   _____

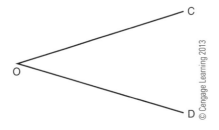

3. How many degrees are in a right angle?   _____

4. Which two triangles below are congruent?   _____

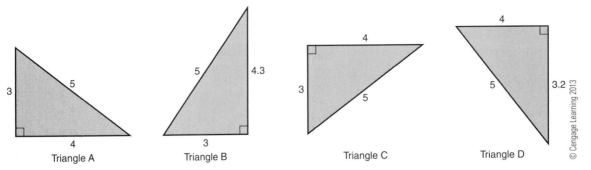

5. Classify by name each of the three triangles below.

    Triangle A _____
    Triangle B _____
    Triangle C _____

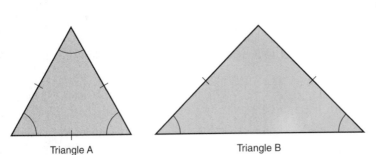 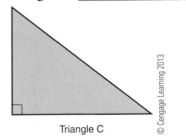

6. Two angles of a triangle have measures of 37° and 53°. What is the measure of the third angle?

     _____

7. The base angles in an isosceles triangle have the same measure. If each base angle measures 62°, what is the measure of the third angle, called the *vertex angle*?

     _____

8. The sum of the measures of two angles in a triangle is 75°. What kind of triangle is this?

     _____

9. The sum of the measures of two angles of a triangle is 90°. What kind of triangle is this?

     _____

10. Tell whether each of the following dimensions form a triangle. If the sides form a triangle, classify what type of triangle it is.

     a. 7, 12, 19   **a.** _____

     b. 6, 8, 10   **b.** _____

     c. 5, 5, 9   **c.** _____

     d. 12.5, 6.7, 12.5   **d.** _____

     e. 15, 15, 15   **e.** _____

11. Answer the following questions about triangle *ABC* below.

     a. What kind of triangle is this?   **a.** _____

     b. What is the perimeter of the triangle?   **b.** _____

     c. Which angle of the triangle has the greatest measure?   **c.** _____

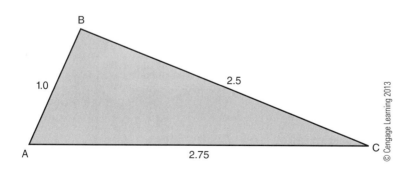

© Cengage Learning 2013

UNIT 21

Polygons and Circles

Basic Principles

Various combinations of polygons and circles make up many of the parts of drafting diagrams and CAD figures. Polygons and circles are closed figures in a plane. Each type of figure separates a plane into three regions: points on the figure, points in the interior of the figure, and points in the exterior of the figure.

Polygons

Three or more line segments that intersect only at their endpoints form polygons. The simplest polygon is a triangle. Polygons that have more than three sides are named according to the number of sides they have. If the number of sides is not known, the figure is known as an *n*-gon, where *n* represents the unknown number of sides. Here is a list of the first eight types of polygons.

Polygon	No. of sides	Polygon	No. of sides
Triangle	3	Heptagon	7
Quadrilateral	4	Octagon	8
Pentagon	5	Nonagon	9
Hexagon	6	Decagon	10

© Cengage Learning 2013

The following are properties of all polygons.

- A polygon has the same number of angles as sides. The vertex of each angle is a *vertex* of the polygon.
- The *perimeter* of a polygon is the sum of the lengths of its sides.

- *Regular polygons* have equal sides and equal angles.
- A *diagonal* of a polygon is a line segment between two non-consecutive vertices.
- The number of diagonals in a polygon equals $\dfrac{n(n-3)}{2}$, where n is the number of sides in the polygon.

Quadrilaterals

Quadrilaterals are four-sided polygons. These include many of the types of figures found in drafting diagrams. A list of the most common quadrilaterals and their basic properties appear below.

Trapezoid	Has exactly one pair of parallel sides called its *bases* 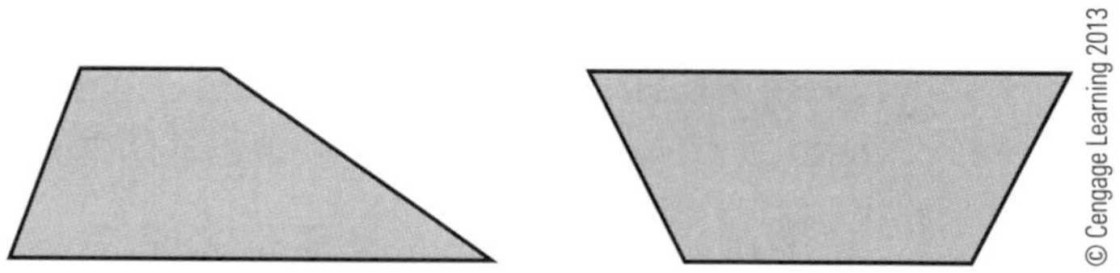
Parallelogram	Has two pairs of opposite parallel and equal sides 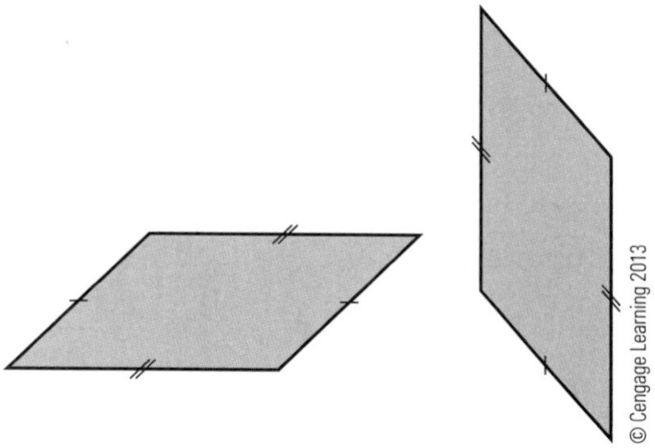

Rectangle	A parallelogram that has four right angles.
Rhombus	A parallelogram that has four equal sides
Square	A rectangle that has four equal sides. NOTE: A square can also be classified as a rhombus that has four right angles.

Circles

A circle is a set of points that are equally distance from a given point called the *center*. The following properties of a circle are important to remember.

- A line segment from the center of the circle to a point on the circle is called a *radius*.
- A line segment through two points on the circle and that passes through the center is called a *diameter*.
- The ratio between the length of the radius of a circle r and the length of its diameter d is $1 : 2$. That is, $d = 2r$ or $r = \dfrac{d}{2}$
- A line through two points of a circle is called a *chord*. A diameter of a circle is the longest chord of the circle.
- The number of degrees in a circle is 360°; the number of degrees in a semicircle is 180°.

- The perimeter of a circle is called its *circumference*. The formulas to calculate the circumference of a circle depend on either r, the length of a radius, or d, the length of a diameter. The formulas are $C = \pi d$ or $C = 2\pi r$, where π (pi) is a number approximately equal to 3.14.
- The basic formula for calculating the area A of a circle is $A = \pi r^2$, where π (pi) is approximately equal to 3.14, and r is the length of the radius of the circle. Another formula used to calculate the area of a circle in this book is $A = 0.7854 \times D^2$, where D is the length of the diameter of the circle. A proof of this formula, where $r = {}^{D}\!/\!{}_{2}$, appears below.

$$A \approx 3.14 \times \left(\frac{D}{2}\right)^2 \approx \frac{3.14}{4} \times D^2 \approx 0.7854D^2$$

Practical Problems

1. What is the perimeter in inches of a square whose sides are each 12 inches?

   _____

2. What is the maximum number of diagonals that can be drawn in an octagon?

   _____

3. The perimeter of a regular pentagon is 135 mm. How long in millimeters is each side of the pentagon?

   _____

4. Draw a nonagon in the space below. Then draw all of its diagonals. What is the maximum number of diagonals in the nonagon?

   _____

5. What is the difference between the perimeters of these two regular octagons?   _____

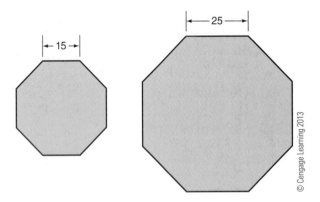

6. What is the perimeter of the regular hexagon below?   _____

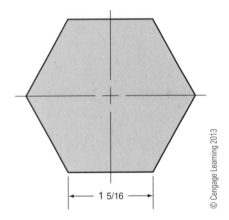

For problems 7 through 14, use 3.14 as the value of π.

7. To the nearest tenth inch, what is the circumference of a circle whose radius is 5.35 inches?   _____

8. What is the radius in millimeters of a circle whose diameter is 116 mm?   _____

9. What is the approximate difference between the circumference of a wheel, whose diameter is 9.25 units long, and a pulley, whose diameter is 8.75 units long? Express the answer to the nearest hundredth unit.   _____

10. A certain collar has a wall thickness of 11.65 mm. The outside diameter is 57.64 mm. What is the inside diameter in millimeters, to the nearest tenth?

_____

11. The circumference of a circle is 135.65 cm. What is the length, in centimeters, of the radius of this circle? Express the answer to the nearest tenth.

_____

12. What is the difference between the circumference of a 3.75-diameter pulley and the circumference of a 9.25-diameter wheel? Express the answer to the nearest hundredth.

_____

CAD Problems

13. What is the difference between the circumferences of the large and small circles in the CAD figure below?

_____

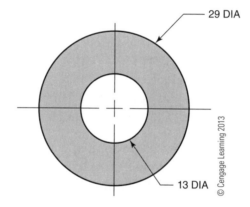

29 DIA

13 DIA

© Cengage Learning 2013

14. How many degrees are between each hole location in this CAD figure? _____

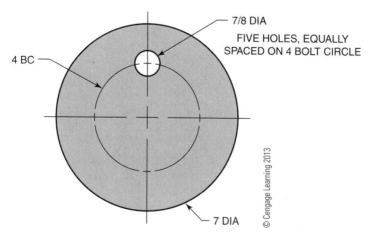

7/8 DIA

FIVE HOLES, EQUALLY SPACED ON 4 BOLT CIRCLE

4 BC

7 DIA

© Cengage Learning 2013

UNIT 22

Constructions

Geometric constructions rely on only two tools: a drawing compass and an unmarked straight-edge. These tools do NOT rely on units of measurement. In theory, the figures that are constructed with these tools are *exact*. Drafting triangles, rulers, and protractors are NOT construction tools. These are measurement tools that provide approximations of exact figures.

- A drawing compass is a hinged device that typically has two arms. One arm has a sharpened point at its end, and the other arm has a pencil at its end. Using a compass, you can draw curved lines and circles, or measure and reproduce equal distances.
- A straight-edge is an unmarked ruler. It enables you to draw the line between two given points or to create a set of *collinear* points; that is, points that lie on the same line.

Although today's drafters and CAD engineers have more sophisticated tools and software at their disposal, it is important for them to know how to use these two basic construction tools as they lay out their work and visualize the operations to be performed.

EXAMPLE: Bisect segment *AB* using a compass. Show all construction lines.

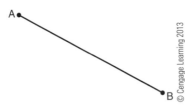

STEP 1: Place the compass point at point *A*. Spread the arms of the compass to choose a radius that is greater than half $\overline{AB}$. Then draw the circle using that radius.

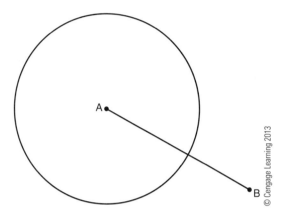

© Cengage Learning 2013

STEP 2: Without changing the distance between the arms of the compass, place the compass point at *B* and draw a second circle. The two circles intersect in two points.

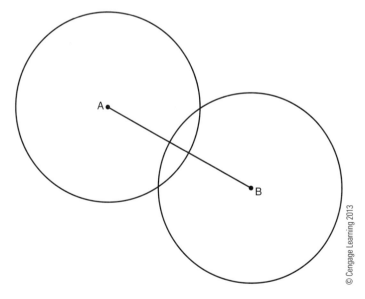

© Cengage Learning 2013

STEP 3: Use a straight-edge or ruler and connect the points of intersection between the circles. The point of intersection on $\overline{AB}$ is the midpoint of $\overline{AB}$, the point that bisects it.

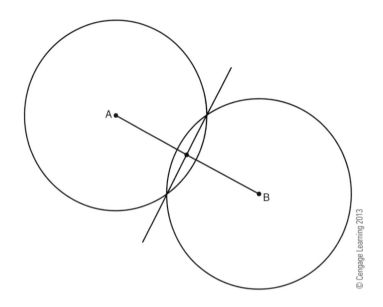

© Cengage Learning 2013

Practical Problems

Use a compass and straight-edge (ruler) to complete the following constructions. If you need more space, copy the given figures on a sheet of paper and complete the constructions there.

1. Bisect arc **CD**. (Hint: Draw segment **CD**. The bisector of $\overline{CD}$ also bisects $\overset{\frown}{CD}$.)

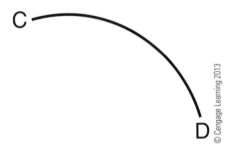

© Cengage Learning 2013

2. Bisect ∠**AOB**. (Hint: Construct a circle with *O* as center. From each point of intersection on *OA* and *OB*, construct circles having the same radius.)

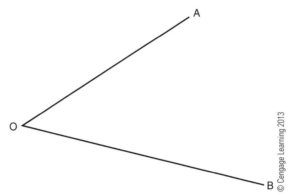

3. Divide line **CD** into seven equal parts. (Hint: Draw a line at an acute angle down from **C** and use the same distance to mark off seven equal segments on that line. Then connect point *D* with the seventh mark on the line and construct six parallel segments from each mark to *CD*.)

4. Construct a perpendicular line from point **L** to line **MN**.

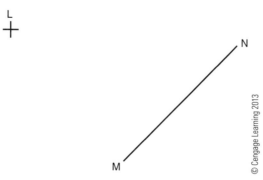

5. a. A *tangent* to a circle is a line that is perpendicular to a radius at its endpoint on the circle. Construct the tangent to the circle at point **T**. At point **T**, construct a 1-inch segment that has **T** as its midpoint.

   b. Construct the tangent from **P** to the lower part of the circle and label the point of tangency.

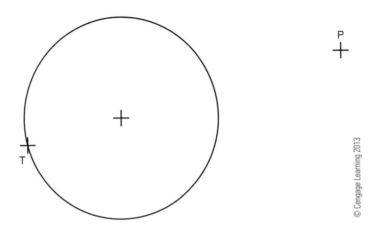

6. Inscribe a regular hexagon within a 84-mm diameter circle.

7. Construct the square that has AB as one of its diagonals. (Hint: The diagonals of a square are equal and each is the perpendicular bisector of the other.)

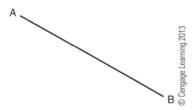

© Cengage Learning 2013

8. The distance *across the flats* is the distance across any set of parallel lines. Construct a regular hexagon using line **CD** as the across the flats distance.

© Cengage Learning 2013

9. Construct an equilateral triangle with line **EF** as a base.

© Cengage Learning 2013

10. The distance *across the corners* is the distance from the vertex of one pair of sides to the vertex of the opposite pair of sides. Construct a regular octagon with line **GH** as the across the corners distance.

G ————————————— H   © Cengage Learning 2013

11. Follow the steps below to construct a circle through points *A*, *B*, and *C*.

STEP 1:   Connect **A** and **B**; connect **B** and **C**.

STEP 2:   Bisect $\overline{AB}$; bisect $\overline{BC}$.

STEP 3:   Place the compass point at the intersection of the bisectors and use the distance to **A** as a radius to draw the circle.

A
+

+ C

B +

© Cengage Learning 2013

12. Follow the steps below to construct an arc tangent to line **AB** at **B,** and
that passes through point **C.**

STEP **1:** Draw $\overline{BC}$.

STEP **2:** Construct the perpendicular bisector of $\overline{BC}$.

STEP **3:** Construct the line perpendicular to $\overline{BC}$ at **B**.

STEP **4:** Use the intersection of these lines as the center of a circle. Use the distance
to **B** as the radius. Draw the arc.

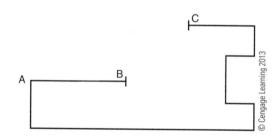

© Cengage Learning 2013

13. Construct a square using **CD** as a side.

© Cengage Learning 2013

14. Follow the steps below to construct triangle **ABC** using $\overline{AB}$ as the base. Make **AC** = 45 mm and **BC** = 89 mm.

STEP 1:   Use 45 mm as a radius and draw an arc from point A.

STEP 2:   Use 89 mm as a radius and draw an arc from point B intersecting the arc from Step 1.

STEP 3:   Label the point of intersection between the arcs as **C**.

STEP 4:   Draw $\overline{AC}$ and $\overline{BC}$.

NOTE: The following problems are more difficult. They require the application of many basic constructions.

15. In the CAD figure below, follow the given steps to construct a 25.40-mm diameter hole equidistant from centers **A** and **B** and equidistant from centers **C** and **D**.

STEP 1:   Connect **A** and **B** and then construct the perpendicular bisector of AB.

STEP 2:   Connect **C** and **D** and then construct the perpendicular bisector of CD.

STEP 3:   Use the intersection of these two perpendicular lines as the center of a circle. Use the compass to draw the circle and add a label showing its dimension in mm.

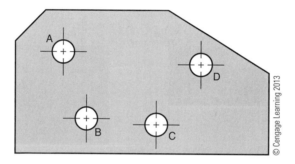

© Cengage Learning 2013

16. On a separate sheet of paper, use full scale to lay out this plate. (Do not use a protractor.)

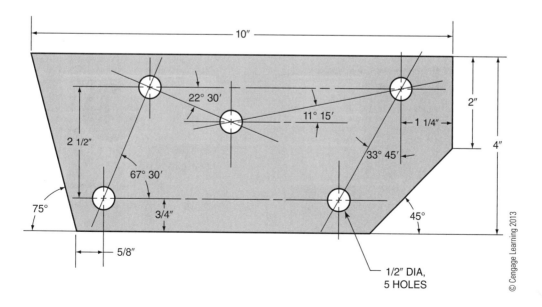

© Cengage Learning 2013

17. Use a compass and a ruler marked in inches to lay out sets of holes on the CAD drawing below.

STEP 1: Bisect ∠**B** and construct three equally spaced ¼" diameter holes. Make the first hole ¾" from point **B** and the last hole 2⁵⁄₁₆" from point **B.**

STEP 2: Bisect ∠**A** and construct four equally spaced ⅜" diameter holes. Make the first hole ⅞" from point **A** and the last hole 3⁹⁄₁₆" from point **A.**

STEP 3: Construct a ¾" diameter circle 1³⁄₁₆" from point **C** and 1⅜" from point **D.** Label each of the dimensions on the construction.

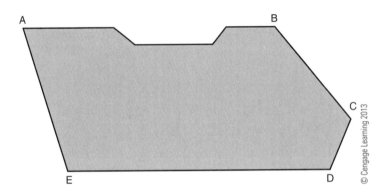

© Cengage Learning 2013

18. Use the given figures below to construct an arc of radius 1¼ inches that is also tangent to the given arc and to the given line. Show all construction lines and label the points of tangency.

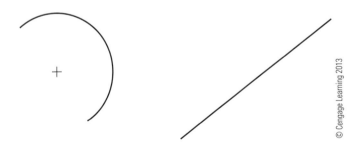

© Cengage Learning 2013

19. Use the figures below to construct an arc of radius 42 mm that is also tangent to the two given arcs. Show all construction lines and label the points of tangency.

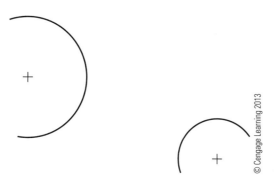

© Cengage Learning 2013

20. Use the figures below to construct an arc of radius 1 inch that is also tangent to the inside of arc **A** and the outside of arc **B**. Show all construction lines and label the points of tangency.

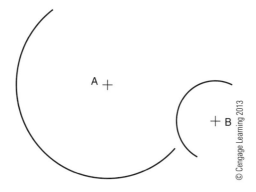

© Cengage Learning 2013

21. The CAD drawing below, labeled COMPLETE FIGURE, shows all its
dimensions. Construct an *exact* copy of this complete figure using
the reference points and segments that appear in the figure labeled
INCOMPLETE FIGURE. Use the *exact* dimensions from the complete
figure, and mark the points of tangency where directed by your
instructor.

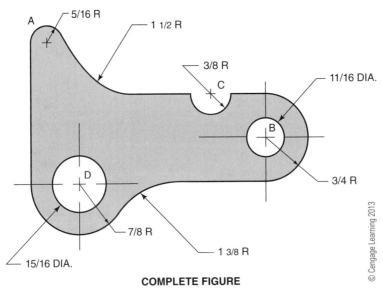

COMPLETE FIGURE

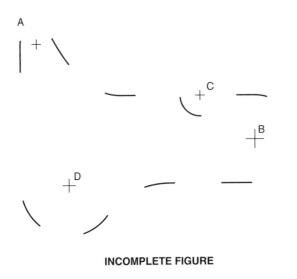

INCOMPLETE FIGURE

SECTION

Measurement

UNIT 23

Linear Measure

Basic Principles

Linear measure refers to "straight line measurement." A drafter must often express given units in larger or smaller units when laying out and dimensioning drawings. When the drawing is being made from an actual object, micrometers and Vernier calipers are often used. The use of all drafting equipment requires accuracy and precision. The two systems of measurements commonly in use are the U.S. customary system and the metric system. Common linear units in the U.S. system and the metric system are given below.

COMMON U.S. LINEAR UNITS	
12 inches (in.)	= 1 foot (ft.)
3 feet (ft.)	= 1 yard (yd.)
16½ ft.	= 1 rod
5,280 ft.	= 1 mile (mi.)

© Cengage Learning 2013

COMMON METRIC LINEAR UNITS	
1 millimeter (mm)	= 0.001 meter (m)
1 centimeter (cm)	= 0.01 m
1 decimeter (dm)	= 0.1 m
1 meter (m)	= 1.0 m

© Cengage Learning 2013

Converting Feet and Inches to Inches

When converting feet and inches to inches, multiply the number of feet by 12 in./ft., and then add the number of inches to the product.

EXAMPLE: Express 3 feet 7¾ inches in inches. (3′−7¾″)

STEP 1: Multiply 3 ft. by 12 in./ft.

3 ft. × 12 in./ft. = 36 in.

NOTE: The feet units cancel out in the calculation leaving inches as the remaining unit.

STEP 2:   Add 36 inches and 7¾ inches

$$36 \text{ in.} + 7\frac{3}{4} \text{ in.} = 43\frac{3}{4} \text{ in.}$$

$$\text{So, } 3'\text{--}7\frac{3}{4}'' = 43\frac{3}{4} \text{ in.}$$

Converting Inches to Feet and Inches

To express inches in terms of feet and inches, divide the number of inches by 12 in./ft. to find the number of feet. The remainder is the number of inches.

EXAMPLE: Express 145½ inches in feet and inches.

STEP 1:   Divide 145 in. by 12 in./ft. The quotient is in feet, and the remainder is in inches.

$$145 \text{ in.} \div \frac{12 \text{ in.}}{1 \text{ ft.}} = 145 \text{ in.} \times \frac{1 \text{ ft.}}{12 \text{ in.}}$$

NOTE: The inch units cancel out in the calculation leaving the units in feet.

$$
\begin{array}{r}
12 \quad \longleftarrow \text{(Quotient in feet)} \\
12\overline{)145\tfrac{1}{2}} \\
-12 \\
\overline{25} \\
-24 \\
\overline{1} \quad \longleftarrow \text{(Remainder in inches)}
\end{array}
$$

STEP 2:   Add 1 inch to the given ½ inch.

$$1 \text{ in.} + \frac{1}{2} \text{ in.} = 1\frac{1}{2} \text{ in.}$$

STEP 3:   Express the quotient and the remainder in feet and inches.

$$12 \text{ ft. } 1\frac{1}{2} \text{ in. } = 12' - 1\frac{1}{2}''$$

$$\text{So, } 145\frac{1}{2}'' = 12' - 1\frac{1}{2}''.$$

NOTE: To convert feet and inches into a decimal in inches, find the total number of inches in the measurement and use mental math, TABLE II, FRACTION AND DECIMAL EQUIVALENTS on page 393, or a calculator to convert this number to an exact or approximate decimal value.

EXAMPLE: Express 2 feet 7½ inches as a decimal to the nearest hundredth.

STEP 1:   Multiply 2 feet and 12 in./ft. The product is in inches.

2 ft. × 12 in./ft. = 24 in.

STEP 2:   ½ = 0.5, so 7½ inches in decimal form equals 7.5 inches

STEP 3:   Add the inches: 24 in. + 7.5 in. = 31.5 in.

$$\text{So, } 2 \text{ ft. } 7\frac{1}{2} \text{ in. } = 31.5 \text{ in.}$$

Skill Problems

1. Express each measurement in inches.

    a. 7 ft. _____   b. 3 ft. 11 in. _____

    c. 5 ft. 7¾ in. _____   d. 6 ft. 3¼ in. _____

2. Express each measurement in feet and inches.

    a. 9.25 yd. _____   b. 9½ yd. _____

    c. 11.75 yd. _____   d. 6⅓ yd. _____

3. Express each measurement in inches rounded to the nearest hundredth as necessary.

   a. 3 ft. 9 in. _____

   b. 4 ft. 11¼ in. _____

   c. 5 ft. 4⁷⁄₁₆ in. _____

   d. 11 ft. 8⅜ in. _____

4. Express the readings indicated on this scale, to the nearest eighth or sixteenth inch.

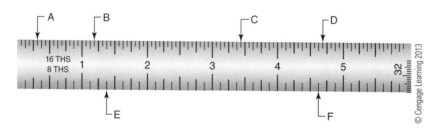

© Cengage Learning 2013

A _____

B _____

C _____

D _____

E _____

F _____

5. Express the readings indicated on this scale to the nearest 32nd or 64th inch.

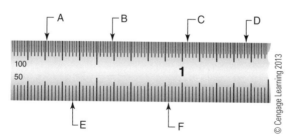

© Cengage Learning 2013

A _____

B _____

C _____

D _____

E _____

F _____

6. Express the readings indicated on this scale to the nearest 50th or 100th inch.

© Cengage Learning 2013

A _____

B _____

C _____

D _____

E _____

F _____

7. Express the readings on this metric scale to the nearest millimeter.

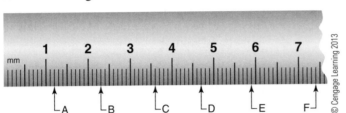

A _____

B _____

C _____

D _____

E _____

F _____

© Cengage Learning 2013

8. Study the examples below that show readings of .857″ and .263″ on a pair of micrometer scales. Then read the settings on the drawings of the six 0.001-inch micrometer scales below the examples and record the settings in parts A through F that follow.

EXAMPLES:

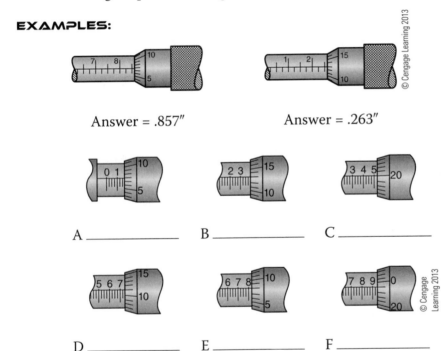

Answer = .857″

Answer = .263″

A _____   B _____   C _____

D _____   E _____   F _____

© Cengage Learning 2013

9. Study the example below that shows a reading of 2.359″ on a Vernier scale. Then read and record in inches the Vernier caliper measurements on the three caliper scales labeled **A**, **B**, and **C** that follow.

EXAMPLE:

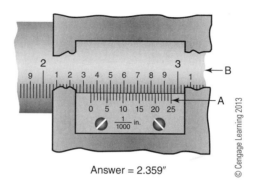

Answer = 2.359″

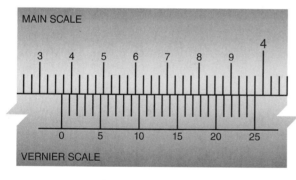

A _____

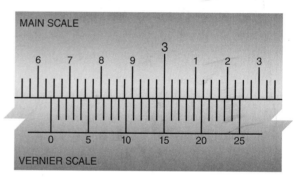

B _____

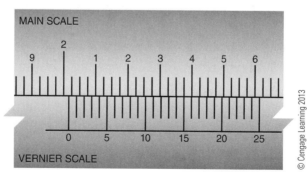

C _____

10. Determine the six readings on the ¼" architect's scale below.

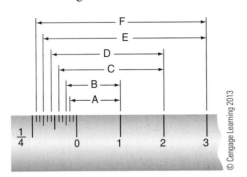

A _____

B _____

C _____

D _____

E _____

F _____

11. Determine the four readings on the ⅜" architect's scale below.

A _____

B _____

C _____

D _____

12. Determine the six readings on the ¾" architect's scale below.

A _____

B _____

C _____

D _____

E _____

F _____

© Cengage Learning 2013

13. Measure each line using the indicated scale and record its length in the labeled spaces labeled **A** through **N** provided below. Be as accurate as the specific scale will allow.

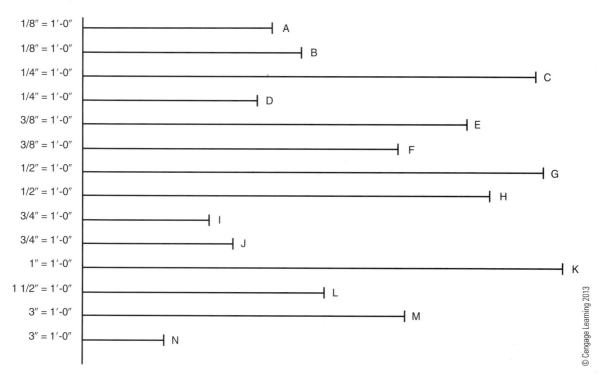

A. _____   B. _____   C. _____   D. _____

E. _____   F. _____   G. _____   H. _____

I. _____   J. _____   K. _____   L. _____

M. _____   N. _____

14. Using an engineer's scale, measure each line and report its measure in feet to the nearest tenth.

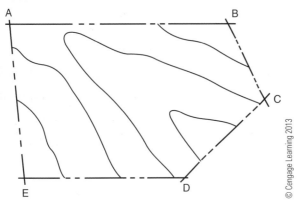

AB when 1″ = 10.0′ _____

BC when 1″ = 20.0′ _____

CD when 1″ = 30.0′ _____

DE when 1″ = 40.0′ _____

EA when 1″ = 50.0′ _____

© Cengage Learning 2013

15. Use the CAD drawing below and the given dimensions to calculate the perimeters of the deck, bedroom, and living room. Express the answer in feet and inches.

Deck _____

Bedroom _____

Living room _____

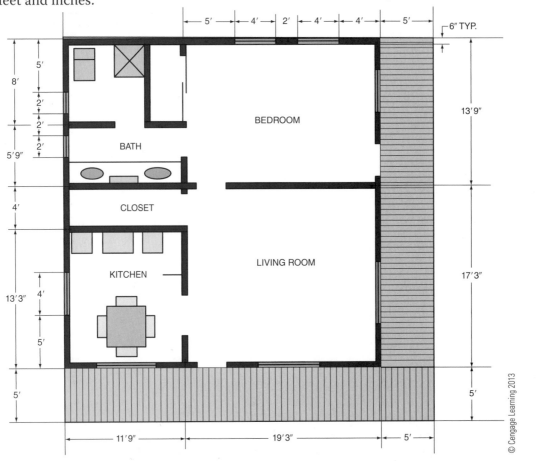

© Cengage Learning 2013

UNIT 24

Basic Principles

Drafters must often draw parts that are too large or too small to be drawn full scale. They must then choose and convert actual measurements to a proper scale. If an actual part is large, the dimensions for a drawing are reduced by multiplying the actual dimensions of the part by a scale factor less than 1. If the actual part is small, the dimensions for the drawing are increased by multiplying the actual dimensions of the part by a scale factor greater than 1. This operation permits easy reading of a drawing.

Skill Problems

In problems 1 to 8, use full scale to measure the lengths of the given lines to the indicated degree of accuracy.

1. Measure these lengths to the nearest ¼".

a ⊢————————————————————————┤   © Cengage Learning 2013   Line a _____

b ⊢—————————————————————┤   Line b _____

2. Measure these lengths to the nearest ⅛".

a ⊢—————————┤   © Cengage Learning 2013

b ⊢———————————————————————┤   Line a _____

                                                             Line b _____

3. Measure these lengths to the nearest ¹⁄₁₆″.

a ⊢————————————⊣   Line a _____

b ⊢——————————————————⊣   Line b _____

4. Measure these lengths to the nearest millimeter.

a ⊢————————————⊣   Line a _____

b ⊢——————————————————⊣   Line b _____

5. Measure these lengths to the nearest ¹⁄₆₄″.

a ⊢—————————⊣   Line a _____

b ⊢——————————————⊣   Line b _____

6. Measure these lengths to the nearest centimeter.

a ⊢———————————————⊣   Line a _____

b ⊢——————————⊣   Line b _____

7. Measure these lengths to the nearest ¹⁄₃₂″.

a ⊢——————————————————⊣   Line a _____

b ⊢————————⊣   Line b _____

8. Measure these lengths to the nearest millimeter.

a ⊢————————————————⊣   Line a _____

b ⊢———————————⊣   Line b _____

CAD Problems

9. On this CAD drawing, a scale of ¼ inch represents 1 inch on the actual part. Measure each labeled dimension **A** through **F**, to the nearest ⅛", and determine its actual length.

A _____

B _____

C _____

D _____

E _____

F _____

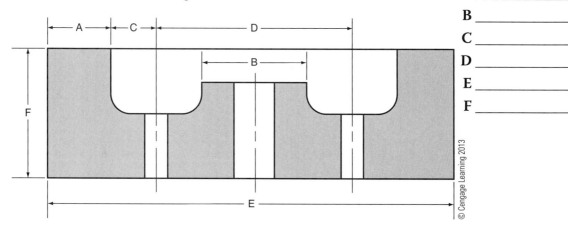

© Cengage Learning 2013

10. Measure each labeled dimension **A** through **F**, to the nearest 1⁄16", and determine its actual length in inches. The scale is ½" = 1".

A _____

B _____

C _____

D _____

E _____

F _____

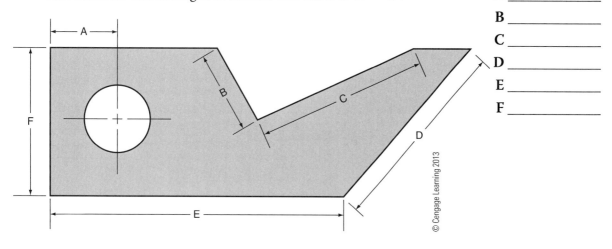

© Cengage Learning 2013

11. Measure the labeled dimensions A through H on this full-scale CAD drawing and record their lengths to the nearest millimeter.

A _____

B _____

C _____

D _____

E _____

F _____

G _____

H _____

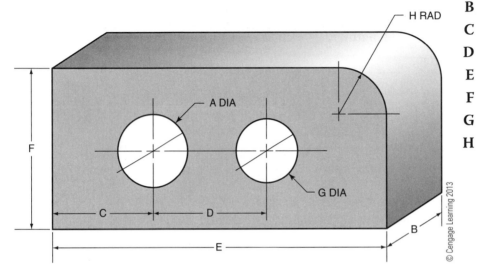

© Cengage Learning 2013

12. On this CAD drawing, a scale of ⅜" represents 1" on the part. Measure the labeled dimensions A through F and record their lengths to the nearest ⅛".

A _____

B _____

C _____

D _____

E _____

F _____

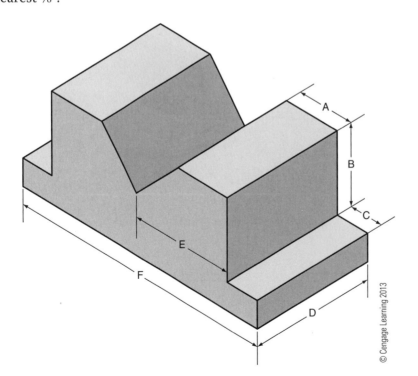

© Cengage Learning 2013

13. This CAD drawing was made with the scale ¾″ = 1″. What are the
dimensions of parts A through F to the nearest ⅛″?

A _____

B _____

C _____

D _____

E _____

F _____

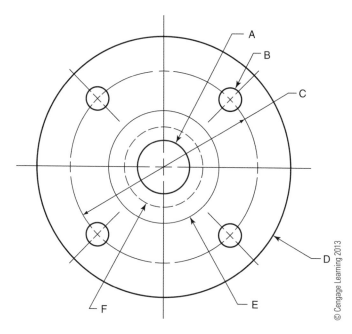

© Cengage Learning 2013

14. Measure lengths A through F in this CAD drawing to the nearest
centimeter.

A _____

B _____

C _____

D _____

E _____

F _____

© Cengage Learning 2013

15. Use the indicated scale for this CAD drawing of the plot plan below. Then measure and record the lengths in feet of dimensions A through G.

A _____

B _____

C _____

D _____

E _____

F _____

G _____

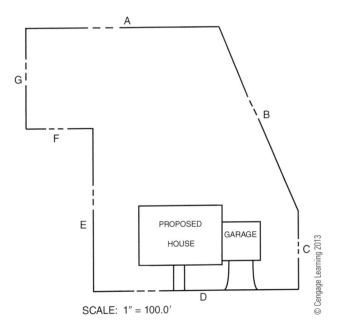

SCALE: 1″ = 100.0′

16. Use the scale indicated for the CAD drawing below and measure and record the lengths in feet of dimensions A through L.

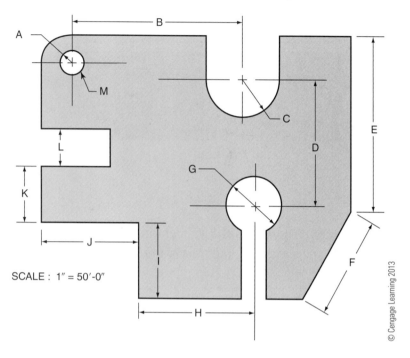

SCALE : 1″ = 50′-0″

© Cengage Learning 2013

A. _____   B. _____   C. _____   D. _____

E. _____   F. _____   G. _____   H. _____

I. _____   J. _____   K. _____   L. _____

M. _____

UNIT 25

Area

Basic Principles

Area is a property of all closed two-dimensional objects. Calculating area involves multiplication, and as a result, an area measure is always expressed in *square units,* such as square inches (in.²) or square centimeters (cm²).

NOTE: Before multiplying linear dimensions to calculate an area, be sure that the linear measures have the same unit.

You can use the formulas below to calculate A, the area of each of the following common two-dimensional figures.

Triangle

$$A = \frac{1}{2}bh$$

b = a base (side) of the triangle

h = the height (altitude) to that base

Rectangle

$$A = bh$$

b = a base (length)

h = the height to that base (width)

Square

$$A = s^2$$

s = side

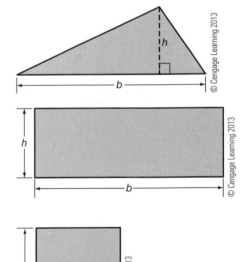

© Cengage Learning 2013

© Cengage Learning 2013

© Cengage Learning 2013

Trapezoid

$$A = \frac{1}{2}h\,(b + B)$$ h = the height (altitude)

b = the shorter base

B = the longer base

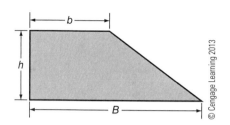

Circle

$A = \pi r^2$ π (pi) $\approx$ 3.1416

$A \approx .7854D^2$ r = radius

D = diameter

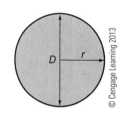

Parallelogram

$A = bh$ b = a base (side)

h = the height (altitude)
to that base

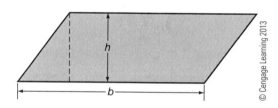

Areas of Polygons

To help you solve these problems, use reference TABLE I, "Equivalent U.S. Customary and Metric Units of Measure," in Section II of the Appendix on page TBD.

Practical Problems

1. How many square inches are in a 12″ by 18″ rectangular sheet of drafting vellum?

    _____

2. What is the surface area, in square units, of the faces **A** and **B** on this CAD drawing of a set-up block?

    A _____

    B _____

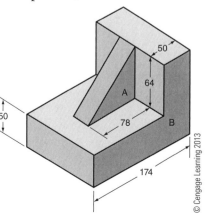

3. How many square inches are in a square with 7.53-inch sides? Round the answer to the nearest hundredth.

_____

4. Calculate, in square feet, the area of the rectangular floor of a room that shows measures of 21'-0" × 16'-0" on a blueprint.

_____

5. What is the area of this parallelogram in square units?

_____

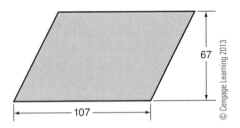

6. Calculate the area of the trapezoidal floor in this CAD diagram. Express the answer in square units.

_____

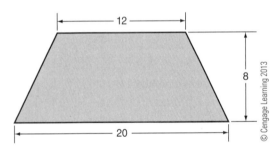

7. What is the area, of the triangle inscribed in this circle to the nearest thousandth?

_____

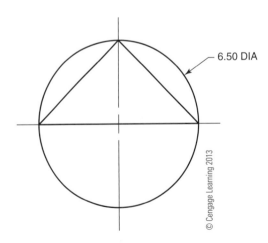

8. What is the difference, in square inches, between the areas of a 9″ × 12″ rectangular piece of drafting vellum and an 18″ × 24″ rectangular piece of vellum?

_____

9. Calculate, in square units, the area of the triangle inscribed in this circle.

_____

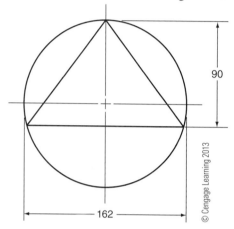

90

162

© Cengage Learning 2013

10. The weight of 20-gauge black iron is 1.5 lb. per sq. ft. Calculate the weight in pounds of a rectangular piece of 20-gauge black iron measuring 48″ × 96″.

_____

CAD Problems

11. What is the area of this drawing of a drafter's 30°–60°–90° triangle minus the cutout? Express the answer to the nearest thousandth.

_____

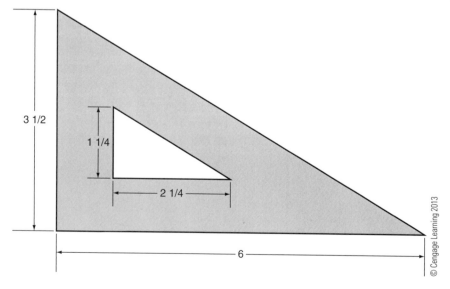

3 1/2

1 1/4

2 1/4

6

© Cengage Learning 2013

12. A rectangular piece of sheet metal is punched as shown below. Calculate the area remaining after the punching operation. Express the answer to the nearest hundredth.   _____

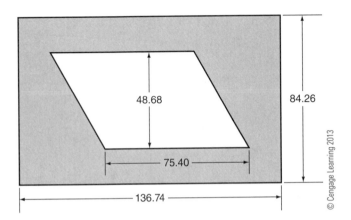

13. The drawing of a triangular steel plate below has a square hole in it. Calculate the area of the triangular plate minus the area of the hole.   _____

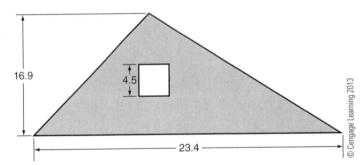

14. What is the area, in square feet, of the drawing of the side elevation of this house? Do not include the area of the 4-foot square window in the calculation.   _____

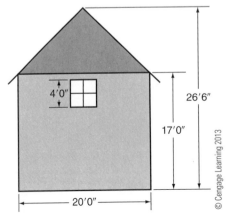

15. How much weight is saved by cutting a rectangular hole in the CAD drawing of the 22-gauge black iron shim below? The dimensions of the hole are expressed in inches, and the weight of the 22-gauge black iron shim is 1.25 lb. per sq. ft.

_____

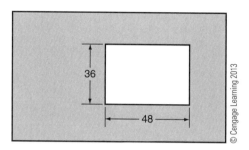

16. What is the total area of this CAD drawing of a transition piece? Express the answer to the nearest thousandth.

_____

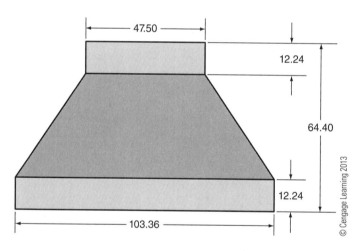

17. A steel block has a rectangular angular groove milled through its upper surface. What is the area of this milled groove? Round the answer to the nearest thousandth.

_____

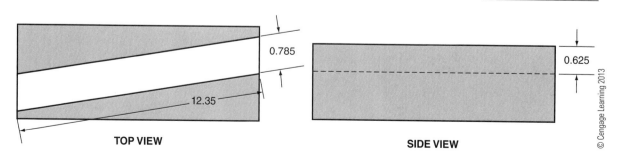

TOP VIEW   SIDE VIEW

18. Calculate the total area of this CAD drawing of a template.   _____

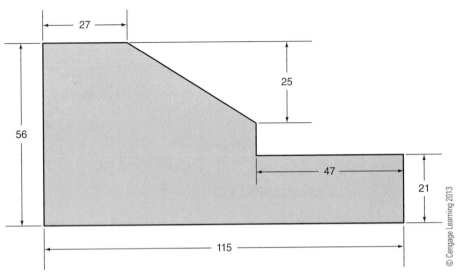

19. Calculate the total area of this CAD drawing of a hopper piece. Express
     the answer to the nearest hundredth.   _____

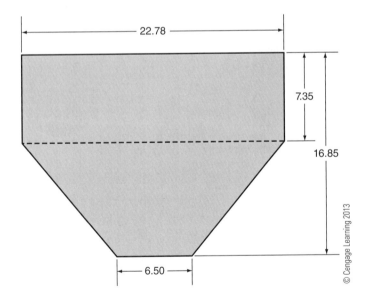

20. Calculate the amount of waste in this CAD drawing of a rectangular sheet metal piece. Express the answer to the nearest hundredth.

_____

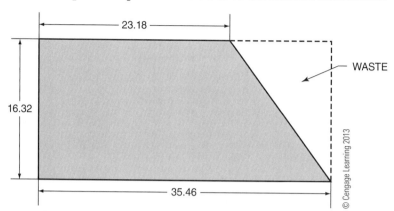

21. Using the CAD drawing below, determine the areas in square feet of the vacation cabin and the deck. Express the answers as decimals in hundredths.

Cabin _____

Deck _____

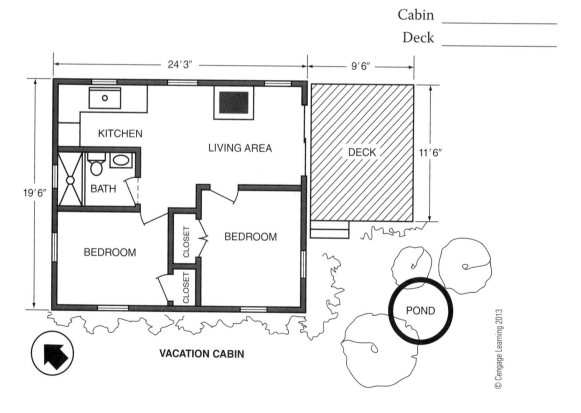

22. Use the CAD drawing below and calculate the area in square units of this object. _____

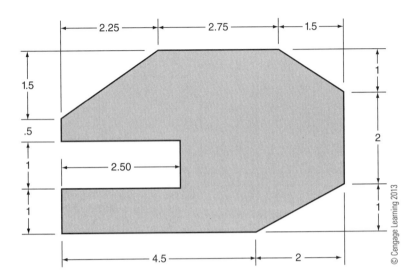

23. Use the CAD drawing of the house below and determine the total area in square feet of all the windows. _____

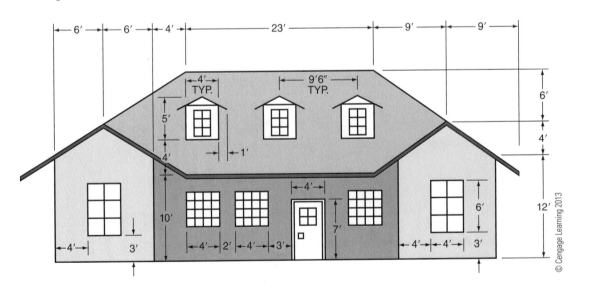

24. Determine the area in square units of the CAD drawing of a spacer below.

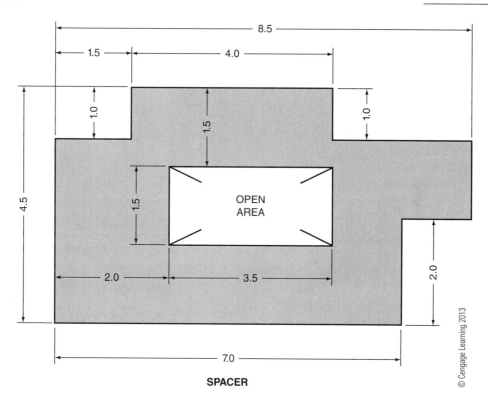

SPACER

25. Determine the total area in square feet of the parallelogram-shaped subdivisions of land in the drawing below. Then, calculate the area in square feet of each building lot in the subdivisions **A** through **D**.

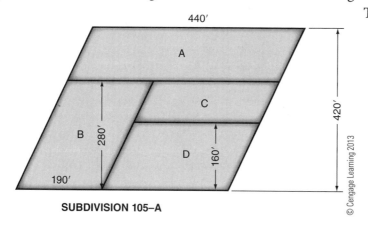

SUBDIVISION 105–A

Total area _____

A _____

B _____

C _____

D _____

26. Calculate the total area in square feet of the parallelogramed-shaped building lots shown in the drawing below. Then calculate the area of each building lot A through D.

Total area _____

A _____

B _____

C _____

D _____

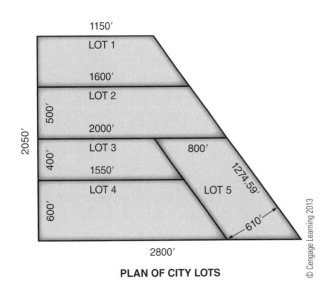

27. Calculate the total area in square feet of the five city building lots in the diagram below. Then calculate the area of each lot, 1 through 5.

Total area _____

Lot 1 _____

Lot 2 _____

Lot 3 _____

Lot 4 _____

Lot 5 _____

PLAN OF CITY LOTS

© Cengage Learning 2013

Areas of Circles and Cylinders

Practical Problems

28. Use 3.142 as the value of π and calculate, to the nearest hundredth inch, the diameter of a circle whose area is 188 square inches.

_____

29. Use 3.1416 as the value of π and calculate, to the nearest thousandth square foot, the area of a circular table top with a diameter of 9 feet.

_____

30. A *sector* of a circle is a pie-shaped region of the circle. What is the area of the sector removed from this sheet metal disc? Express the answer to the nearest thousandth square unit. (Hint: The missing area is $^{40°}\!/_{360°}$ or ⅑ of a circle.)

_____

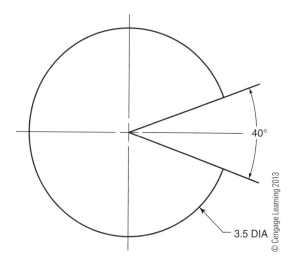

40°

3.5 DIA

© Cengage Learning 2013

31. What is the area of a washer with an inside diameter of 1¾ inches and an outside diameter of 3⅝ inches? Express the answer to the nearest thousandth square inch.

_____

32. A stamping made of a circular 22-gauge tin plate weighs 1.263 lb. per sq. ft. If the area is 78 square inches, what is its weight? Round the answer to the nearest thousandth pound.

_____

33. Calculate the total surface area of a right circular cylinder whose base has a diameter of 4.10 inches, and whose height is 8.5 inches. Express the answer to the nearest hundredth square inch.

_____

CAD Problems

34. Calculate, to the nearest hundredth, the area of this CAD drawing of a shim, minus the areas of the two holes.

Area of template _____

% Waste _____

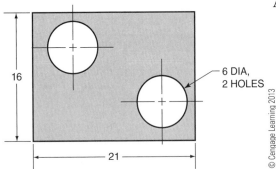

6 DIA,
2 HOLES

16

21

© Cengage Learning 2013

35. Use this CAD drawing of a stamping and calculate its area to the nearest hundredth square unit.

_____

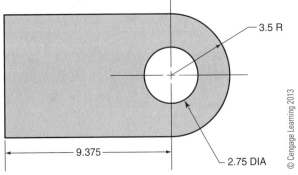

3.5 R

9.375

2.75 DIA

© Cengage Learning 2013

36. What is the area of this CAD figure of a brass gasket to the nearest thousandth square unit?

_____

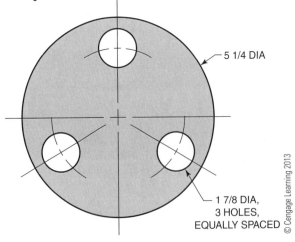

5 1/4 DIA

1 7/8 DIA,
3 HOLES,
EQUALLY SPACED

© Cengage Learning 2013

37. Calculate, in square units, the area of this circular ring.

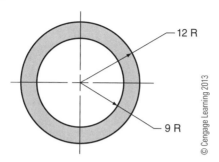

© Cengage Learning 2013

38. What is the area of this CAD drawing of a shim? Express the answer to the nearest thousandth square unit.

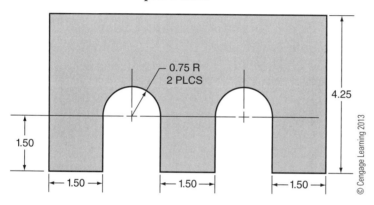

© Cengage Learning 2013

39. This CAD figure of a special cover is made from 30-gauge tin plate that weighs 0.491 lb. per sq. ft., and its dimensions are expressed in inches below. Calculate, to the nearest hundredth, the weight in pounds of the cover.

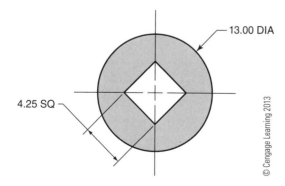

© Cengage Learning 2013

40. What is the area of this CAD figure of a connector link? Express the
    answer to the nearest hundredth square unit. _____

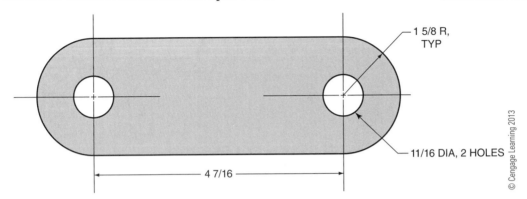

41. Calculate the area of the darkest shaded portion of this washer. Express
    the answer to the nearest thousandth square unit. _____

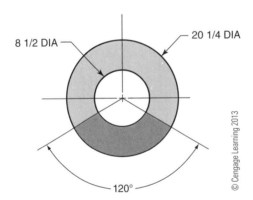

42. How many square units of waste are there in this CAD drawing of a
    sheet metal stamping? Express the answer to the nearest thousandth. _____

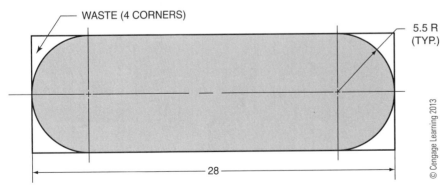

43. Calculate the area of this circle after the triangular piece is removed. Express the answer to the nearest hundredth square unit. _____

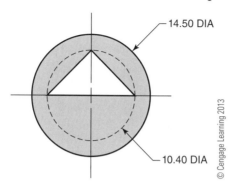

14.50 DIA

10.40 DIA

© Cengage Learning 2013

44. What is the lateral (outside) surface area to the nearest thousandth square unit of this stepped shaft? _____

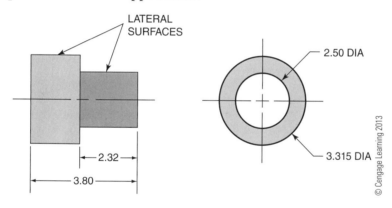

LATERAL SURFACES

2.50 DIA

3.315 DIA

2.32

3.80

© Cengage Learning 2013

45. Calculate, to the nearest thousandth square unit, the area of this template after the three semicircles have been removed. Then calculate to the nearest whole number the percent waste due to the removal of these punchings.

Area of template _____

% Waste _____

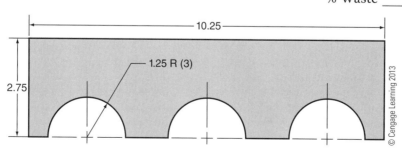

10.25

1.25 R (3)

2.75

© Cengage Learning 2013

46. What is the area of this shim to the nearest thousandth square unit? _____

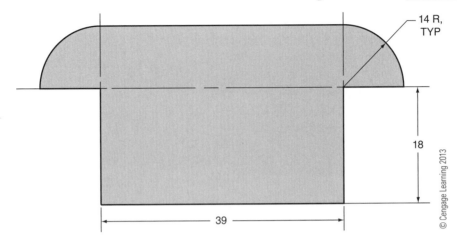

47. What is the total area of four templates like the one shown below? Express the answer to the nearest thousandth square unit. _____

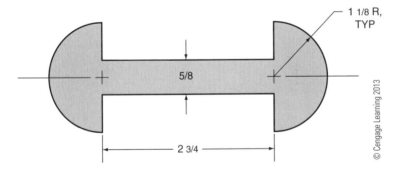

48. Use the CAD drawing below and calculate the area in square units of the circular object minus the areas of the holes. _____

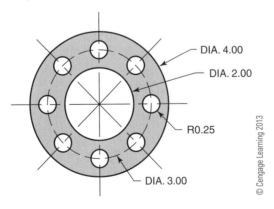

49. All holes on this CAD drawing have a ⅝" diameter. Calculate to the nearest square inch the area of the object minus the areas of the holes. _____

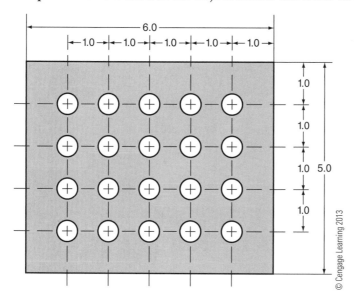

50. Use the CAD drawing of the special gasket below and calculate its area to the nearest ten thousandths square units. _____

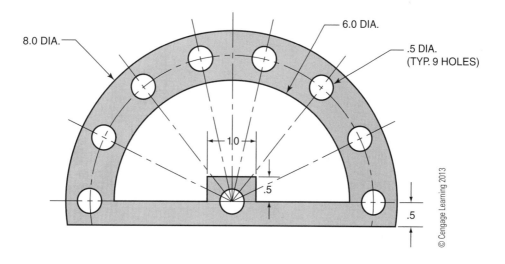

51. Use the CAD drawing of the symmetrical gasket below to determine its surface area in square units. _____

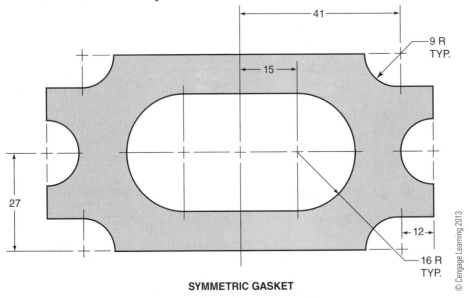

SYMMETRIC GASKET

© Cengage Learning 2013

52. Determine the surface area in square inches of the front and back sides of the CAD drawing of a symmetrical spacer below. Round the answer to three decimal places. _____

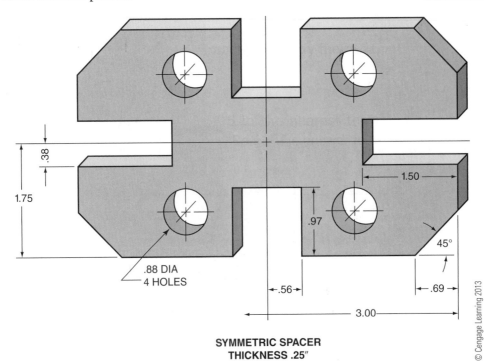

SYMMETRIC SPACER
THICKNESS .25″

© Cengage Learning 2013

UNIT 26

Volume

Basic Principles

Volume is the amount of space occupied by a body. Volume is a three-dimensional property of all real objects. The following are examples of common three-dimensional objects and the formulas for calculating their volumes V.

Rectangular Prism

A *rectangular prism* has 6 *faces*, 12 *edges*, and 8 *vertices*.

- A face is a flat surface. In a rectangular prism, the faces are rectangles.
- An edge is the intersection of two faces.
- A vertex is a point where edges intersect.

In a rectangular prism, the opposite faces are both rectangular and congruent.

A *square prism* is a rectangular prism whose opposite faces are squares that are not all congruent to each other.

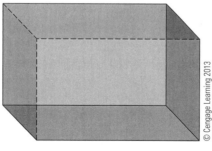

$$V = L \times H \times W \qquad \begin{aligned} L &= \text{length} \\ H &= \text{height} \\ W &= \text{width} \end{aligned}$$

Cube

A cube is a square prism whose six faces are congruent squares.

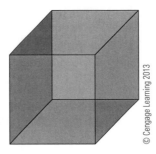

$$V = e^3 \qquad e = \text{length of each edge}$$

Square-Based Pyramid

A pyramid has triangular faces that converge at a single point called the *vertex*. The *base* of a pyramid can be any polygonal shape, such as a triangle, rectangle, square, and so on. As its name implies, a *square-based* pyramid has a square base.

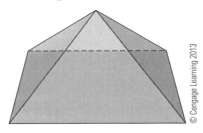

$$V = \frac{Bh}{3} \qquad B = \text{area of its square base } (B = s^2)$$

$$h = \text{height of the pyramid}$$

Right Circular Cylinder

A right circular cylinder has two congruent, circular bases. The distance between the bases is the altitude, h of the cylinder.

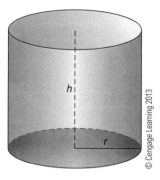

$$V = \pi r^2 h$$ r = radius of the circular bases
h = height of the cylinder

Right Circular Cone

A right circular cone is similar to a pyramid except that its *lateral surface* is rounded, not flat. Similar to a pyramid, the lateral surface of a cone converges at a vertex. The base of a circular cone is a circle with radius r.

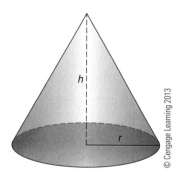

$$V = \frac{\pi r^2 h}{3}$$ r = radius of the circular base
h = height of the cone

Weight

In many instances, drafters and engineers are concerned with the weights of shafts and other common machine parts. The weight of a machine part or the combined weight of a product might affect the operation of the product as intended. Thus, volume calculations are extremely important. The total weight of an object equals the product of its volume and the weight per unit of volume.

Volumes of Prisms and Pyramids

Use reference TABLE I, "Equivalent U.S. Customary and Metric Units of Measure," in Section II of the Appendix on page TBD to help you solve the following problems.

Practice Problems

1. How many cubic inches are in 6 cubic feet?   _____

2. How many cubic feet are in 7 cubic yards?   _____

3. Express 6 cubic yards, 13 cubic feet in cubic feet alone.   _____

4. What is the volume, in cubic inches, of a 3-foot square based prism?   _____

5. Calculate, to the nearest cubic foot, the volume of a rectangular-shaped room 12 ft. 6 in. wide $\times$ 20 ft. 6 in. long $\times$ 8 ft. 3 in. high.   _____

6. A rectangular prism has edges that are 9 inches and a volume of 640 cubic inches. Calculate the depth in inches.   _____

7. A container in the shape of a rectangular prism measures 51 in. $\times$ 74 in. $\times$ 48 in. The volume of material to be stored in this container is 231 cubic inches per gallon. How many gallons, to the nearest hundredth, can this container hold?   _____

8. A rectangular prism with height 38.5 in. and length 25.25 in. has a 57.0-gallon capacity. What is the width of the prism to the nearest hundredth inch? (Hint: Rewrite the volume formula solved for W:
    $$W = \frac{V}{L \times H}.)$$ _____

9. A piece of wood measures 17 in. $\times$ 9 in. and has a volume of 512 in.^{3} How many inches thick is the wood? (Hint: Thickness = Width, so
    $$W = \frac{V}{L \times H}.)$$ _____

10. A specific type of steel weighs 0.283 pounds per cubic inch. Calculate the weight in pounds of a rectangular steel bar whose height and width are 3″, and whose length is 7.5″. Express the answer to the nearest hundredth.

_____

11. Determine the number of cubic yards of concrete in a pyramid with a square base measuring 8 yards on each side and an altitude (height) of 42 feet. Express the answer to the nearest hundredth.

_____

12. Calculate the difference between the volumes, in cubic yards, between a cube with 3-foot sides and a cube with 8-foot sides. Express the answer to the nearest hundredth.

_____

13. A rectangular piece of steel 2-feet long and 2-feet wide weighs 490 pounds. Steel weighs 490 pounds per cubic foot. What is the height of this piece of steel to the nearest ¼-foot?

_____

14. What is the volume, in cubic inches, of a piece of wood whose dimensions are ¾ in. × 5 in. × 12 in.?

_____

CAD Problems

15. Calculate the volume in cubic units of this CAD drawing of a step block.

_____

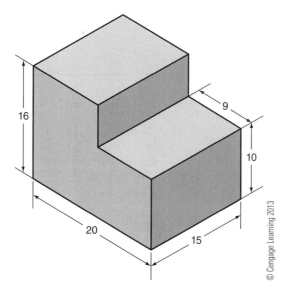

16. What is the volume in cubic inches of this CAD drawing of a
    rectangular core box?   _____

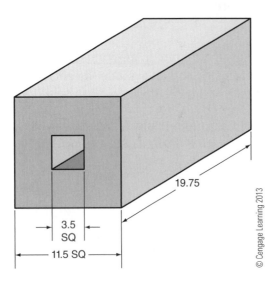

17. Calculate the volume of this CAD drawing of a notch block to the
    nearest thousandth.   _____

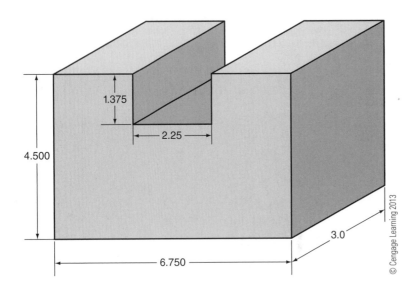

18. Calculate, to the nearest hundredth, the volume in cubic units of this square-based pyramid.

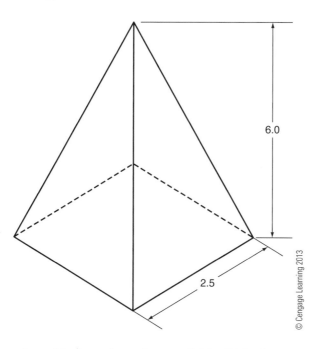

6.0

2.5

© Cengage Learning 2013

19. Calculate, in cubic feet, the volume of this CAD drawing of a wall.

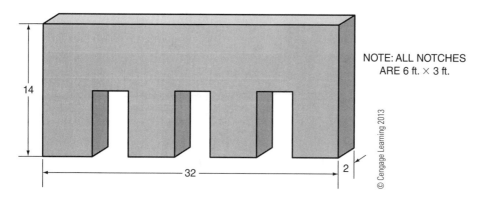

NOTE: ALL NOTCHES
ARE 6 ft. × 3 ft.

14

32

2

© Cengage Learning 2013

20. Determine the volume of this CAD drawing of a step step block to the nearest thousandth.   _____

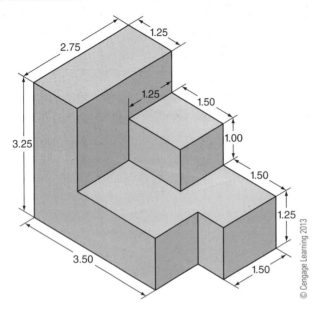

© Cengage Learning 2013

Volumes of Cylinders and Cones

Use reference TABLE I on page TBD to help you solve the following problems.

Practical Problems

21. Calculate the volume of a cylinder that is 8 inches long and whose circular base has a diameter of 3.25 inches. Express the answer to the nearest thousandth.   _____

22. A cone is 7.5 feet high, and its circularbase has a diameter of 2.5 feet. What is the volume, in cubic feet, of the cone to the nearest thousandth?   _____

23. What is the volume, in cubic feet, of a circular cylinder 18 inches in diameter and 12 feet long? Express the answer to the nearest hundredth.   _____

24. Calculate, in cubic yards, the volume of a cone 25 feet in diameter and 18 feet high. Express the answer to the nearest hundredth.   _____

25. A cylindrical oil storage-tank is 8 ft. in diameter and 15 ft. high. What is its volume, in cubic feet, to the nearest hundredth?

_____

26. Calculate the gallon capacity of a cylindrical oil drainage pan 30 inches in diameter and 8 inches high. Express the answer to the nearest thousandth.

_____

27. Calculate the volume of a cone whose base has a 24 mm diameter and whose height is 19 mm. Express the answer to the nearest thousandth.

_____

28. A steel collar has an inside diameter of 1⅛ in., an outside diameter of 3¾ in., and a length of 2⁷⁄₁₆ in. The steel used to make the collar weighs 0.283 pounds per cubic inch. Calculate, to the nearest hundredth, the weight in pounds of the collar.

_____

29. A 6-inch long piece of bar stock is turned on a lathe from a 5½-inch diameter to a 3¼-inch diameter. The bar stock remains cylindrical after the machining operation. How many cubic inches of metal are removed? Express the answer to the nearest hundredth.

_____

30. The weight of the bar stock used in a bar is 0.283 pounds per cubic inch. Calculate, in pounds, the weight of a piece of bar stock that has a 1½" diameter and is 4½" long. Express the answer to the nearest hundredth.

_____

CAD Problems

31. Calculate to the nearest thousandth the volume of this centering pin in the CAD drawings below.

_____

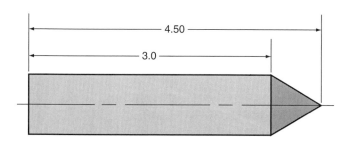

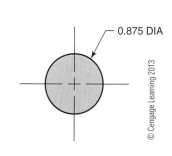

0.875 DIA

© Cengage Learning 2013

32. A 2-inch square hole is machined in the CAD drawings of the cylinder below. Calculate, to the nearest thousandth, the remaining volume, in cubic inches, of the cylinder after machining the square hole.

_____

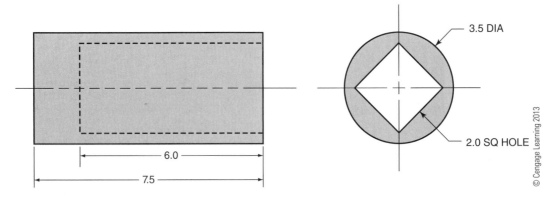

3.5 DIA

2.0 SQ HOLE

6.0

7.5

© Cengage Learning 2013

33. What is the volume of this CAD drawing of a holding block to the nearest hundredth?

_____

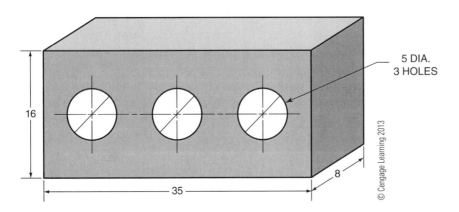

5 DIA.
3 HOLES

16

35

8

© Cengage Learning 2013

34. Calculate, to the nearest thousandth, the volume of this CAD drawing of a centering pin.

_____

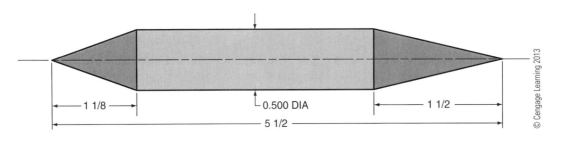

1 1/8

0.500 DIA

1 1/2

5 1/2

© Cengage Learning 2013

35. Bronze weighs 3.2 pounds per cubic inch. Calculate, to the nearest hundredth, the weight in pounds of this bronze bushing. All dimensions on CAD drawings of the bushing are in inches.

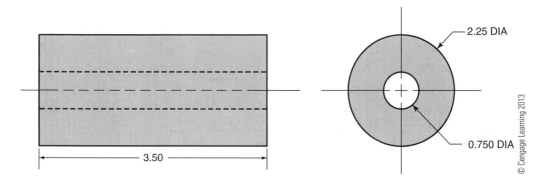

2.25 DIA

0.750 DIA

3.50

36. A flat-bottomed hole is milled in this retainer cap. The hole is 7 cm in diameter and 20 cm deep. What is the volume in cubic centimeters of the cap after the milling operation? Express the answer to the nearest thousandth.

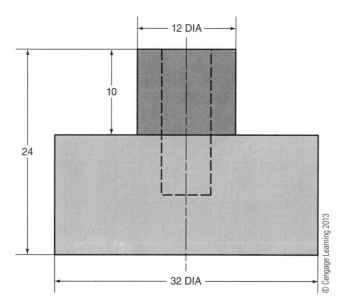

12 DIA

10

24

32 DIA

37. The *end portion* of this Lathe Live Center includes its conical and cylindrical portions only. What is the volume of the end portion of this Live Center to the nearest thousandth?

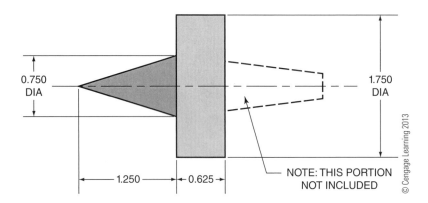

38. Brass weighs 3.0 pounds per cubic inch. Calculate, to the nearest hundredth, the weight in pounds of the brass coupling in the CAD drawings below. All dimensions on the brass coupling are in inches.

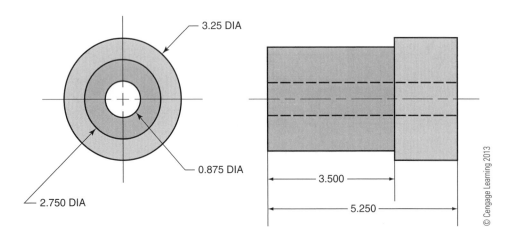

39. What is the volume, to the nearest thousandth, of the special bearing cap in the CAD drawings below? _____

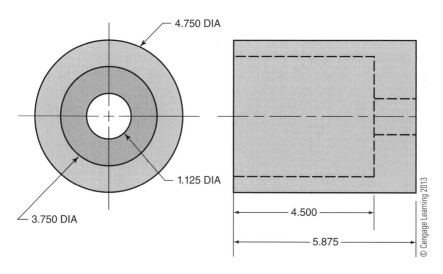

4.750 DIA

1.125 DIA

3.750 DIA

4.500

5.875

© Cengage Learning 2013

40. Calculate, to the nearest thousandth, the volume of this CAD drawing of a hold-down plate. _____

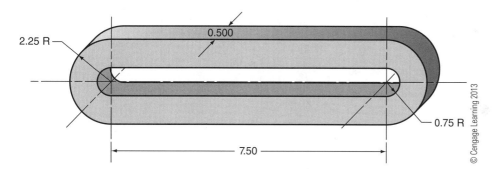

2.25 R

0.500

0.75 R

7.50

© Cengage Learning 2013

41. Use these CAD drawings of this face plate to calculate its volume. _____

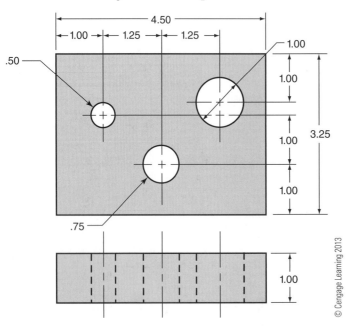

42. Use the CAD drawing below to calculate the weight in pounds of the object. The stock weighs 0.225 pounds per cubic inch and all dimensions are in inches. Express the answer to the nearest hundredth. _____

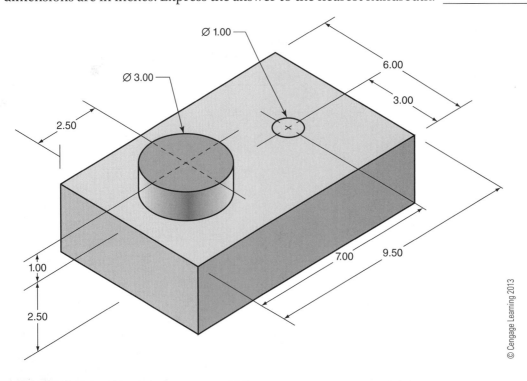

43. Use the CAD drawing below to determine the volume of the object. Express the answer to the nearest hundredth. _____

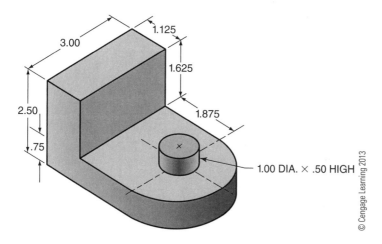

1.125
3.00
1.625
2.50
1.875
.75
1.00 DIA. × .50 HIGH
© Cengage Learning 2013

44. Use the CAD drawings below to calculate the weight in pounds of the object. The stock weighs 0.312 pounds per cubic inch and all dimensions are in inches. Express the answer to the nearest hundredth. _____

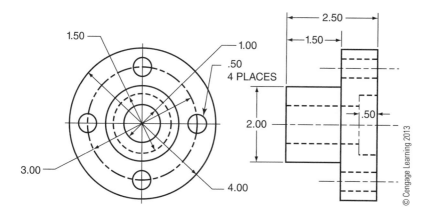

1.50
1.00
.50
4 PLACES
2.50
1.50
2.00
.50
3.00
4.00
© Cengage Learning 2013

45. Use the CAD drawing below to determine the volume of this object.
Express the answer to the nearest hundredth.   _____

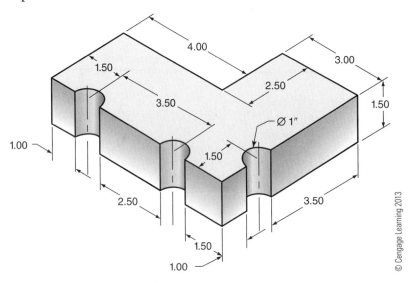

46. Use the CAD drawings below to find the weight in pounds of this object.
The stock weighs 0.425 pounds per cubic inch, and all dimensions are
in inches. Express the answer to the nearest hundredth.   _____

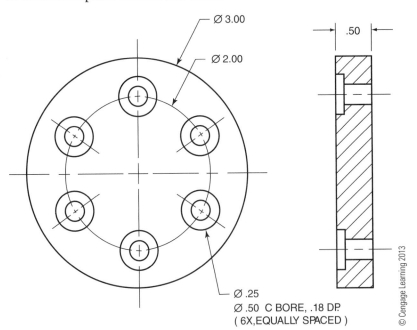

47. Use the CAD drawing below to calculate the weight in pounds of this link if the stock weighs 0.642 pounds per cubic inch. Express the answer to the nearest hundredth.

_____

LINK
SCALE: HALF
THICKNESS: .375 IN.

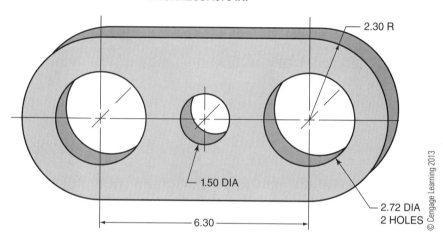

2.30 R

1.50 DIA

6.30

2.72 DIA
2 HOLES

© Cengage Learning 2013

UNIT 27

Equivalent Measurement Units

Basic Principles

There are two major systems of measurement, the U.S. customary system and the metric system. The U.S. customary system is used primarily in the United States, as its name suggests. However, most of the rest of the world uses the metric system. Machines, tooling, and gauges in a factory are often designed for one system of measurement. Architectural and engineering drawings that are dimensioned in one system can be changed into the other system as necessary using given equivalents. Therefore, it is important for drafters to be fluent using either system of measurement and to be able to convert from one to the other.

The metric system is easy to understand if you keep in mind that it is based on powers of 10. The meter is the basic unit of length in the metric system. The prefix of "-meter" tells you what fraction or multiple of a meter is being used. Here are some common prefixes and abbreviations for these units.

METRIC LENGTHS

milli-	=	1/1000 of a meter (mm)	deci-	=	1/10 of a meter (dm)
centi-	=	1/100 of a meter (cm)	kilo-	=	1,000 meters (km)

© Cengage Learning 2013

The following tables contain equivalents between U.S. customary units and metric units of length, area, and volume.

TABLE OF LINEAR EQUIVALENTS

1 in.	=	0.0254 m	39.37 in.	=	1 m
1 in.	=	0.254 dm	3.937 in.	=	1 dm
1 in.	=	2.54 cm	0.394 in.	=	1 cm
1 in.	=	25.40 mm	0.039 in.	=	1 mm
			1 ft. = 0.3048 m		
			1 yd. = 0.9144 m		

© Cengage Learning 2013

TABLE OF AREA EQUIVALENTS

1 in.^2	=	$6,4516 \text{ cm}^2$	$1,550 \text{ in.}^2$	=	1 m^2
1 ft.^2	=	0.0929 m^2	15.50 in.^2	=	1 dm^2
1 yd.^2	=	0.836 m^2	0.155 in.^2	=	1 cm^2
			0.00155 in.^2	=	1 mm^2

© Cengage Learning 2013

TABLE OF VOLUME EQUIVALENTS

1 in.^3	=	16.387 cm^3	$61.023.377 \text{ in.}^3$	=	1 m^3
1 ft.^3	=	0.0283 m^3	61.023 in.^3	=	1 dm^3
1 yd.^3	=	0.7646 m^3	0.061023 in.^3	=	1 cm^3
			0.000059 in.^3	=	1 mm^3

© Cengage Learning 2013

Linear Conversions

Practical Problems

Use the tables of equivalents on the preceding pages or TABLE I in Section II of the Appendix on page 392 to convert the following linear units of measurement as needed. Round all answers to the nearest thousandth as necessary.

1. Express 9 inches in millimeters. _____

2. Express 137 millimeters in inches. _____

3. Express 16½ inches in centimeters. _____

4. Express 4.750 inches in millimeters. _____

5. Express 19 feet in meters. _____

6. Express 7 meters in feet. _____

7. Express 396.35 centimeters in inches. _____

8. Express 23 meters in yards. _____

9. To the nearest thousandth, how many inches equal 17 decimeters? _____

10. To the nearest hundredth, how many inches equal 45 centimeters? _____

11. What is the diameter in millimeters of a 16-inch diameter circle to the nearest tenth in millimeters? _____

12. A 72-inch line is subdivided into eight equal parts. Express the length in millimeters of each subdivision to the nearest tenth. _____

13. A ⁵⁄₁₆″ diameter piece of copper-round stock is 315 centimeters long. What is the maximum number of 7″ studs that can be cut from it? _____

14. A certain collar has a wall thickness of 11.65 mm. The outside diameter is 57.64 mm. Calculate the inside diameter in inches. _____

15. Five line lengths are combined to form one line. The lengths are 2.54 cm, 5.08 cm, 22.225 mm, 6.350 mm, and 0.7015 cm. What is the length in inches of this line to the nearest thousandth? _____

CAD Problems

16. All dimensions on the CAD drawing of the machined plate below are in inches. Express, in millimeters, dimensions **A, B, C,** and **D** on this plate. Round the answers to the nearest thousandth.

A _____

B _____

C _____

D _____

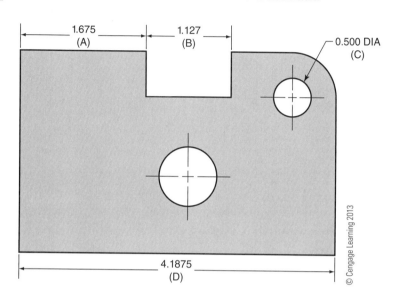

17. The dimensions of this step block are in inches. Calculate the length in centimeters of the missing dimension rounded to the nearest tenth. _____

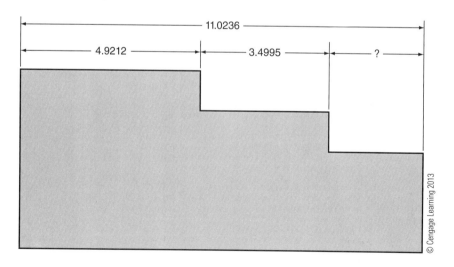

18. The dimensions of this large washer are in millimeters. Determine the wall thickness in inches of the large washer rounded to the nearest thousandth.

_____

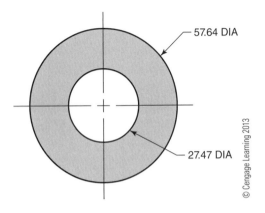

57.64 DIA

27.47 DIA

© Cengage Learning 2013

19. The dimensions of this CAD drawing of a drilled hole are in millimeters.

a. Express the diameter of this hole rounded to the nearest thousandth inch.

a. _____

b. Express the depth of this hole rounded to the nearest thousandth inch.

b. _____

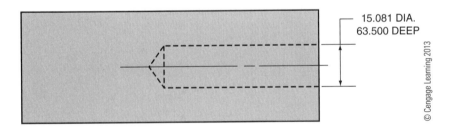

15.081 DIA.
63.500 DEEP

© Cengage Learning 2013

20. The dimensions of this fixture are in inches. Determine dimensions **A** and **B**, expressing each answer in decimeters rounded to the nearest thousandth.

A _____

B _____

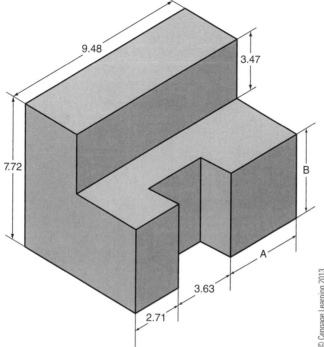

9.48

3.47

7.72

B

A

3.63

2.71

© Cengage Learning 2013

Area Conversions

Practical Problems

Use the tables of equivalents on the preceding pages and TABLE I in Section II of the Appendix on page 392 to convert the following area measurements. Round each answer to the stated specificity.

21. Express 17 square inches rounded to the nearest hundredth square centimeter.

_____

22. Express 939.8 square millimeters in square inches rounded to the nearest hundredth.

_____

23. Express 19 square decimeters in square inches, rounded to the nearest tenth.

_____

24. Express 12.245 square inches to the nearest square centimeter, rounded to the nearest whole number.

_____

25. Express in square centimeters, the approximate area of a circle with a diameter of 1⅞ inches. Express the answer to the nearest ten thousandth. _____

26. A mechanical drafter must calculate the number of square centimeters in the area of a rectangle that is 2¹⁄₁₆ inches long and 1⅜ inches wide. Calculate the area rounded to the nearest hundredth. _____

27. Calculate the number of square inches in the approximate area of a circle with a diameter of 254.00 millimeters. Use 3.14 as the value of π and round the answer to the nearest hundredth. _____

28. A drawing storage room in an engineering firm has a rectangular floor that measures 22 ft. × 36 ft. What is the area in square meters of the floor to the nearest hundredth? _____

29. The area of a floor on a blueprint is 60 square meters. What is its area to the nearest hundredth square yard? _____

30. The floor area of a microfilming room in a small company is 24.5 square meters. What is its area to the nearest hundredth square yard? _____

31. A square piece of galvanized sheet metal has an area of 33.25 square meters. Calculate the area of the sheet metal to the nearest thousandth square yard. _____

32. Calculate the area, in square meters, of a circular table with a diameter of 7 feet. Use 3.14 as the value of π and express the answer to the nearest thousandth. _____

CAD Problems

33. All dimensions on the triangle below are in inches. What is the area of this triangle in square millimeters, expressed to the nearest hundredth? _____

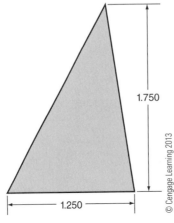

1.750

1.250

© Cengage Learning 2013

34. All dimensions of this trapezoid are in inches. What is its area in square decimeters expressed to the nearest thousandth?

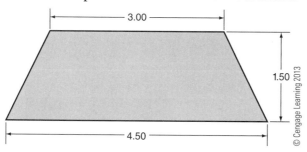

35. All dimensions of this rectangle are in millimeters. Calculate, to the nearest thousandth square inch, the area of the rectangle in square inches.

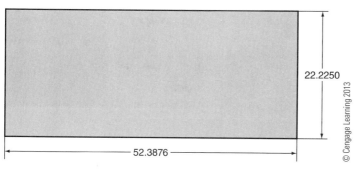

36. All dimensions on this circular ring are in decimeters. Calculate, to the nearest thousandth square inch, the area of the circular ring, using 3.14 as the value of π.

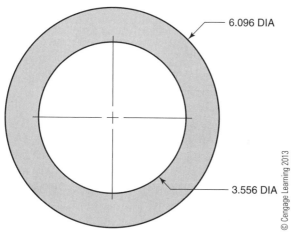

37. All dimensions on this parallelogram are in meters. Calculate, to the nearest hundredth square inch, the area of the parallelogram.

_____

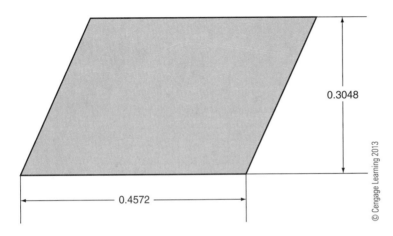

0.3048

0.4572

© Cengage Learning 2013

38. This template has an area of 43.7 square inches. What is its area rounded to the nearest hundredth square centimeter?

_____

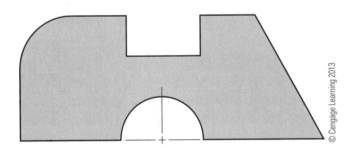

© Cengage Learning 2013

39. This shim has an area of 18.75 square meters. What is its area to the nearest thousandth square foot?

_____

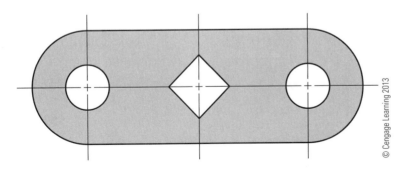

© Cengage Learning 2013

40. This template has an area of 1806.40 square millimeters. What is its area to the nearest hundredth square inch? _____

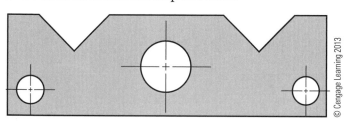

41. All dimensions on this template are in centimeters. Calculate, to the nearest hundredth square inch, the area of the template using 3.14 as the value of π. _____

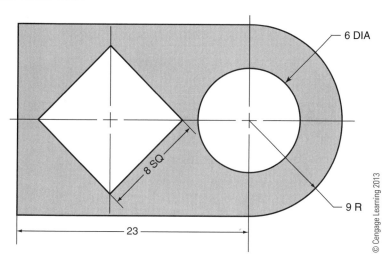

6 DIA

8 SQ

9 R

23

Volume Conversions

Skill Problems

Use the tables of equivalents on the preceding pages or TABLE I in Section II of the Appendix on page 392 to convert the following volume measurements. Round each answer to the desired specificity.

42. Express 3,465 cubic inches to the nearest thousandth cubic meter. _____

43. Express 18 cubic yards to the nearest hundredth cubic meter. _____

44. Express 149 cubic feet to the nearest hundredth cubic meter. _____

45. Express 15.4 cubic decimeters to the nearest hundredth cubic inch. _____

46. Express 3.876 cubic inches to the nearest thousandth cubic centimeter. _____

Practical Problems

47. Calculate the volume, in cubic inches, of a cubical box with 3-decimeter sides. Express the answer to the nearest hundredth. _____

48. A block of steel is 1.8 meters long, 2.6 meters wide, and 1.2 meters high. Calculate, to the nearest hundredth cubic yard, the volume of the block. _____

49. What is the volume, in cubic decimeters, of a cylinder with a radius of 0.75 ft. and a length of 4 ft.? Use 3.14 as the value of π and express the answer to the nearest hundredth. _____

50. The height of a square-based pyramid is 0.95 m, and the sides of its base measure 0.46 m. Calculate, to the nearest thousandth cubic foot, the volume of the pyramid. _____

51. Determine the volume, in cubic centimeters, of a right circular cone with a 5-inch diameter base and an altitude of 13 inches. Use 3.14 as the value of π and express the answer to the nearest hundredth. _____

52. What is the volume in cubic meters of a rectangular drafting room with dimensions 60 ft. × 40 ft. × 9 ft.? Express the answer to the nearest hundredth. _____

53. A cone-shaped pile of sand is 18 ft. high and has a base diameter of 25 ft. Use 3.14 as the value of π and calculate the volume of the pile to the nearest hundredth cubic meter. _____

54. What is the volume, in cubic millimeters, of a rectangular Mylar film eraser that measures 0.250 in. × 1.5 in. × 0.750 in.? Express the answer to the nearest hundredth. _____

CAD Problems

55. All dimensions on this CAD drawing of a gauge are in centimeters. What is the total volume in cubic inches of this gauge, expressed to the nearest hundredth?

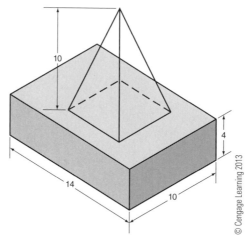

© Cengage Learning 2013

56. All dimensions on the CAD drawings of the steel collar below are in inches. Use 3.14 as the value of π and calculate the volume, in cubic decimeters, of this collar to the nearest ten-thousandth.

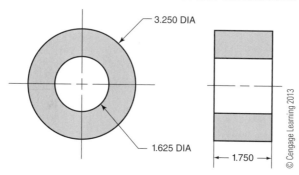

3.250 DIA

1.625 DIA

1.750

© Cengage Learning 2013

57. The dimensions of this CAD drawing of a core box are in yards. Calculate, to the nearest hundredth cubic meter, the volume of the box.

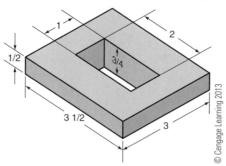

© Cengage Learning 2013

58. All dimensions on the CAD drawing of the spacer below are in centimeters. Use 3.14 as the value of π and calculate, to the nearest hundredth cubic inch, the volume of this spacer. _____

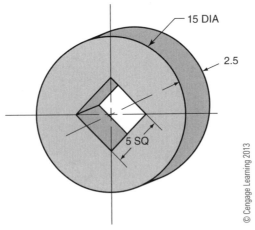

15 DIA

2.5

5 SQ

© Cengage Learning 2013

59. All dimensions on this CAD drawing of a holding block are in meters. Calculate, to the nearest hundredth cubic yard, the volume of the block. _____

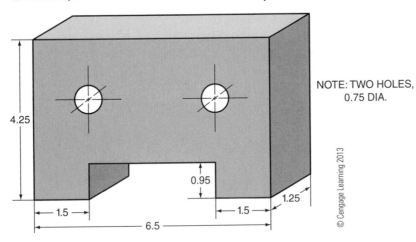

NOTE: TWO HOLES,
0.75 DIA.

4.25

0.95

1.25

1.5

1.5

6.5

© Cengage Learning 2013

60. All dimensions on the CAD drawing below of a gasket are in inches. Calculate the volume, in cubic decimeters, of this gasket. Use 3.14 as the value of π and express the answer to the nearest hundredth. _____

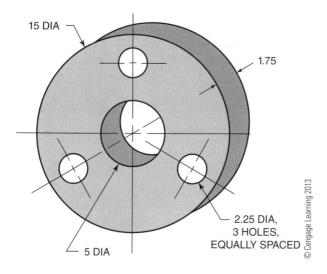

15 DIA

1.75

2.25 DIA,
3 HOLES,
EQUALLY SPACED

5 DIA

© Cengage Learning 2013

61. The levels on this CAD drawing of a step block are 4-cm, 7-cm, and 13-cm square-based prisms that are each 3 cm high. A 1.80-cm diameter hole passes through all three levels. Use 3.14 as the value of π and calculate the volume in cubic inches of this block to the nearest hundredth. _____

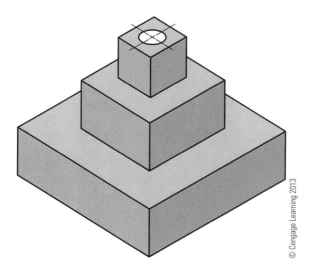

© Cengage Learning 2013

UNIT 28

Angle Measure

Basic Principles

Angles are related to circles, and a circle is a set of points that lie equally distant from a given point. A circle contains 360°. Curved parts of a circle less than 360° are called *arcs* and are also measured in degrees.

The measure of an angle is also based on 360°. An angle whose vertex is at the center of a circle is called a *central angle.* The measure of a central angle is equal to the number of degrees in the arc of the circle between the sides of the angle. In the figure below, the measure of the arc between the sides of the angle is 52°, so the measure of central angle A is also 52°.

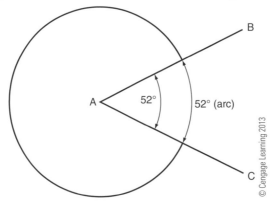

© Cengage Learning 2013

Nearly all types of drawings contain angles, and the size of an opening or a corner on a drawing must be dimensioned. The common units and symbols used to measure an angle in both the U.S. customary system and the metric system are degrees (°), minutes ('), and seconds ("). Common tools used to measure angles include a *protractor*, whose shape is usually a semi-circle. A *drawing compass* is used to draw circles of given radii.

Here are the equivalences between common angle measures.

ANGULAR EQUIVALENTS

1 circle = 360°	1 degree (1°) = 60 minutes (60′)
½ circle = 180°	1 minute (1′) = 60 seconds (60″)
¼ circle = 90°	

© Cengage Learning 2013

Converting Degrees to Minutes

To convert degrees into minutes, multiply the number of degrees by 60 min./degree and then express the product in terms of minutes.

EXAMPLE: Convert 5° to minutes.

$$5 \text{ degrees} \times 60 \text{ min./degree} = 300 \text{ min.}$$
$$5° = 300′$$

NOTE: The degree units cancel out in the calculation.

Converting Minutes to Seconds

To convert minutes to seconds, multiply the number of minutes by 60 sec./min. and then express the product in terms of seconds.

EXAMPLE: Convert 9′ to seconds.

$$9 \text{ min.} \times 60 \text{ sec./min.} = 540 \text{ sec.}$$
$$9′ = 540 \text{ sec.}$$

NOTE: The minute units cancel out in the calculation.

Practical Problems

1. How many degrees are in a semicircle?   _____

2. How many central angles of 15° are in a circle?   _____

3. How many degrees are in four right angles?   _____

4. How many degrees are in ⅛ of a circle?   _____

5. How many degrees and minutes are in angle *X*? (Hint: Measures greater than 60 minutes should be converted to degrees.)

_____

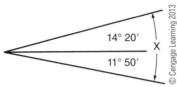

6. How many degrees are in angle *X*?

_____

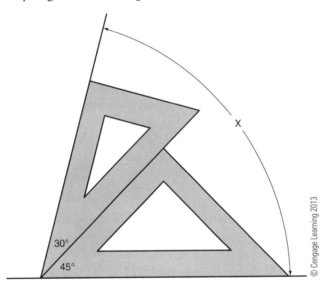

7. The measure of one angle is 8° 15′. What is the total number of degrees and minutes in the sum of the measures of seven of these angles?

_____

8. How many degrees are in angle *X*?

_____

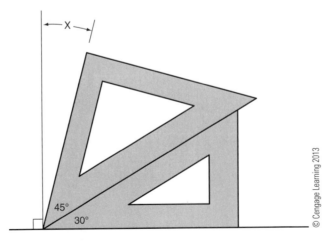

9. Calculate the measure of angle *X* in degrees and minutes.   _____

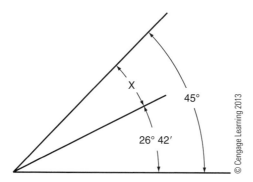

45°

X

26° 42′

© Cengage Learning 2013

10. How many degrees are in a straight angle?   _____

11. Express 900 minutes in degrees.   _____

12. What is the total number of degrees in 840 minutes and 10,800 seconds?   _____

13. Calculate the measure of angle *X* in degrees and minutes.   _____

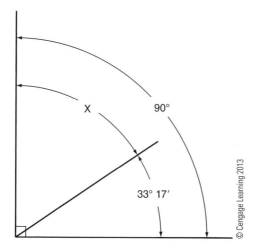

X

90°

33° 17′

© Cengage Learning 2013

14. Calculate the measure of angle *Y* in degrees, minutes, and seconds.   _____

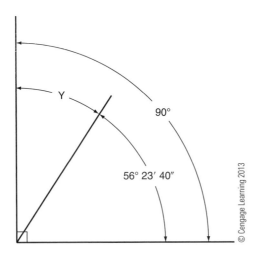

15. The sum of the measures of angle *R* and angle *S* equals 180°. Angle *S* measures 65° 27′. Calculate the measure of angle *R* in degrees and minutes.   _____

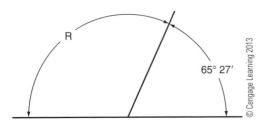

16. The sum of the measures of two angles is 180°. The first angle has a measure of 114° 28′ 37″. Calculate the number of degrees, minutes, and seconds in the second angle.   _____

17. In the figure below, lines **RS** and **TU** intersect to form right angles. Calculate the measures in degrees and minutes, as necessary, of angles **C, E,** and **G.**

∠C _____

∠E _____

∠G _____

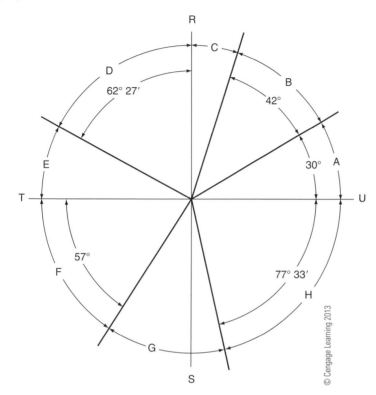

18. An angle measuring 26° 42′ 36″ is divided into four equal parts. Calculate the number of degrees, minutes, and seconds in each part. _____

19. A regular pentagon has five equal sides and five equal angles. Calculate the degree measure of the central angles of the pentagon. _____

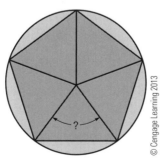

CAD Problems

20. Calculate the number of degrees between the centers of two consecutive holes on this CAD drawing of a base plate. _____

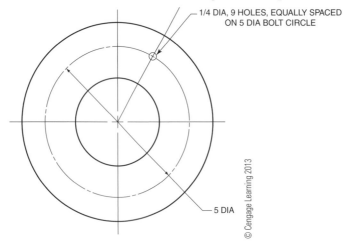

1/4 DIA, 9 HOLES, EQUALLY SPACED ON 5 DIA BOLT CIRCLE

5 DIA

© Cengage Learning 2013

21. A protractor is a measuring tool for angles. Use a protractor and determine the number of degrees in the labeled angles on this CAD drawing of a template.

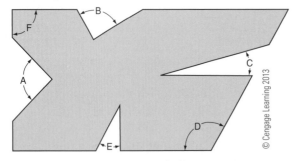

© Cengage Learning 2013

∠A _____
∠B _____
∠C _____
∠D _____
∠E _____
∠F _____

22. The sum of the measures of the interior angles on the drawing of the site plan below equals 360°. Calculate the measure of ∠D in degrees, minutes, and seconds. _____

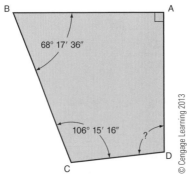

B   A

68° 17′ 36″

106° 15′ 16″

?

C   D

© Cengage Learning 2013

23. Using the CAD drawing below, determine the degree measures of angles **A** through **D**.

∠**A** _____

∠**B** _____

∠**C** _____

∠**D** _____

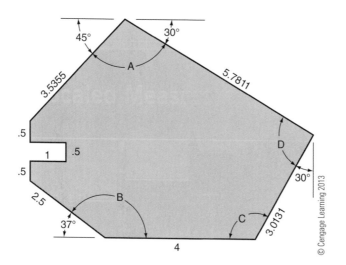

24. Using the CAD drawing below, calculate the measures of angles **C** and **E** in degrees and minutes if the measure of angle **A** is 39° 14′ and the measure of angle **D** is 25° 42′.

∠**C** _____

∠**E** _____

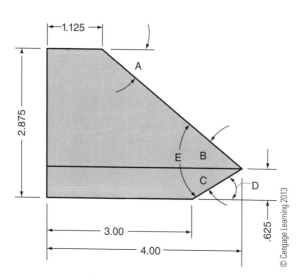

25. Using the CAD drawing below, calculate the measures of angles **C** and **D** in degrees and minutes if the measure of ∠**A** is 22° 18′ and the measure of ∠**B** is 48° 44′.

∠**C** _____

∠**D** _____

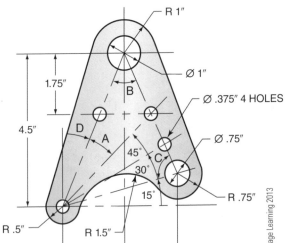

26. Using the CAD drawing below, determine the measures in degrees and minutes of angles **C** and **F** if the measure of ∠**A** is 43° 37′, ∠**B** is 33° 41′, ∠**D** is 25° 16′, and ∠**E** is 66° 21′.

∠**C** _____

∠**F** _____

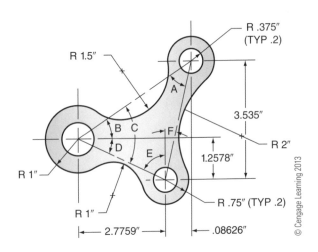

SECTION

Algebra

UNIT 29

Expressions and Equations

Basic Principles

Mathematical expressions and equations are forms of communication. They are used extensively on working drawings as a form of shorthand and as simplified ways to indicate operations that need to be performed. They also are used to identify quantities and units of measurement. Symbols make it possible to create mathematical expressions and write and solve equations.

Expressions can consist of just numbers or combinations of numbers and variables. A *variable* is a letter that usually represents one number. When variables are used to form expressions in combination with numbers, for example, 4**A** or 3**B**, where **A** and **B** might represent dimensions, each expression can be called a *literal term*.

EXAMPLE: The perimeter *P* of this rectangle can be expressed in literal terms, where *L* represents the length of the rectangle, and *W* represents the height.

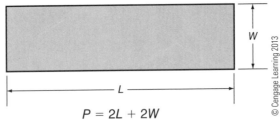

$$P = 2L + 2W$$

Literal terms that have the same variable are called *like terms* and can be combined to form one expression.

EXAMPLE: Find the sum of the like terms.

In this expression, the variable in each term is B. So, these like terms can be combined, starting with the ones in parentheses.

$$1B + 3B + (\underline{7B - 4B}) + (\underline{5B - 3B})$$
$$= \underline{1B + 3B + 3B + 2B}$$
$$= 9B$$

Evaluating Expressions

If you know the value of the variable in an expression, you can *evaluate* (find the value of) the expression.

EXAMPLE: Evaluate the expression $3B$ when B is 15.

Substitute 15 for the variable B in the expression and multiply.

$$3B = 3(15) = 45, \text{ where } 3(15) = 3 \times 15.$$

Solving Equations

Equations are used to solve problems when one (or more) quantities are not known. An *equation* is a mathematical sentence that indicates that two quantities or expressions are equal. The symbol used to indicate equality between two expressions is called the *equals sign* (=). The terms in an equation can include numbers, variables (literal terms), or both numbers and variables.

You can use the **Properties of Equality** to solve equations. Each property is a way to isolate the variable by performing an opposite operation. Study each example below to see how this works.

Addition Property of Equality	Subtraction Property of Equality
If equals are added to equals, the results are equal.	If equals are subtracted from equals, the results are equal.
$$X - 5 = 25$$ $$X - 5 + 5 = 25 + 5$$ $$X = 30$$	$$X + 5 = 25$$ $$X + 5 - 5 = 25 - 5$$ $$X = 20$$
Multiplication Property of Equality	**Division Property of Equality**
If equals are multiplied by equals, the results are equal.	If equals are divided by equals (except by 0), the results are equal.
$$\frac{1}{2}X = 25$$ $$2\left(\frac{1}{2}X\right) = 2(25)$$ $$X = 50$$	$$4X = 16$$ $$\frac{4X}{4} = \frac{16}{4}$$ $$X = 4$$

EXAMPLE: Solve the equation $X + 9 = 17$ for X.

STEP 1:   On the left, 9 is added to the variable X. So, do the opposite and subtract 9 from both sides of the equation to isolate X.

$$X + 9 - 9 = 17 - 9$$

STEP 2:   Simplify both sides of the equation ($X + 0 = X$).

$$X = 8$$

EXAMPLE: Solve the equation $15Y = 45$ for Y.

STEP 1:   The variable X is multiplied by 15. So, do the opposite and divide both sides of the equation by 15.

$$\frac{15Y}{15} = \frac{45}{15}$$

STEP 2:   Simplify both sides of the equation.

$$Y = 3$$

Skill Problems

Evaluate the following expressions for the given values of the variables. Round decimal answers to the nearest hundredth as necessary and write fractional answers in simplest form.

1. $N - 3$ when $N = 9$ _____

2. $5D$ when $D = 4\frac{9}{16}$ _____

3. $\dfrac{S}{3}$ when $s = 3.125$ _____

4. $3x + 2y$ when $x = 4$ and $y = 11\frac{1}{4}$ _____

5. $3(a + b) + 2$ when $a = 6$ and $b = 5$ _____

6. $\dfrac{A - B}{2} + 9$ when $A = 26.32$ and $B = 8.12$ _____

Solve each equation for the given variable.

7. $x + 6 = 52$

8. $26 = 11.6 + y$

9. $8Z - 4Z = 34$ (Hint: Combine like terms first.)

Practical Problems

10. The expression $7x$ represents the length of the shortest side of a triangle. The longest side is 2.9 times the length of the shortest side. What is the length of the longest side in terms of x?

11. The sides of a regular hexagon are equal. If the length of each side is expressed as $9.5H$, what is the perimeter of the hexagon in terms of H?

12. The circumference C of the circular base of a cylinder is 21.87 inches. What is D, the diameter of the base, to the nearest hundredth inch? Use 3.142 as the value of π.

13. A certain screw has 15 threads. This is ⅙ as many threads as on a second screw. If S represents the number of threads on the second screw, what is S?

14. The width W of a rectangle is 24⅜ inches. Its length L is 17¾₁₆ inches longer than its width. What is L in inches?

CAD Problems

15. Express the length of dimension **A** on this strap in terms of dimension **T**.

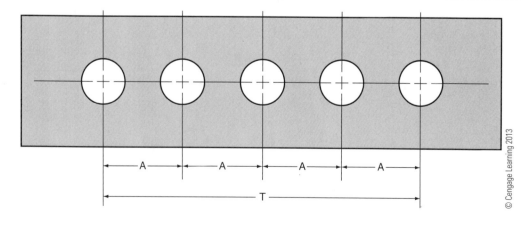

© Cengage Learning 2013

16. Express the length of dimension **L** on this template in terms of
dimension **D**.   _____

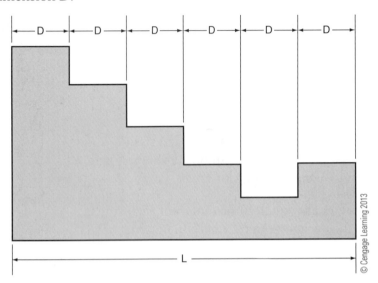

17. The marked lengths in the CAD figure below are expressed in terms of *L*.
Express the lengths of dimensions **A** and **B** in terms of *L*.   **A** _____

B _____

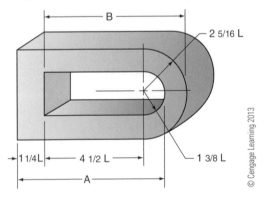

18. The dimensions on the CAD figure below are expressed in terms of
the variable X. Express dimensions **A**, **B**, and **C** in terms of X.

A _____
B _____
C _____

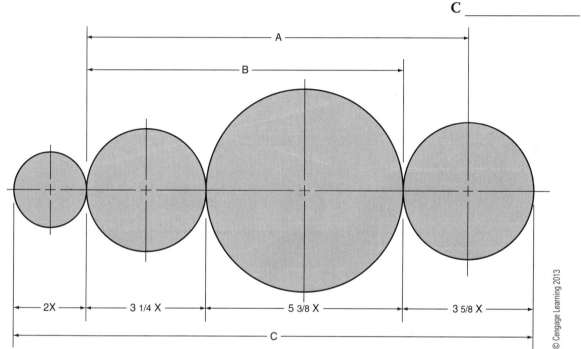

2X   3 1/4 X   5 3/8 X   3 5/8 X

© Cengage Learning 2013

19. Calculate the value of X in the CAD drawing of the bolt and nut below.
Then use the value of X to calculate dimensions **A**, **B**, **C**, and **D**.

X _____
A _____
B _____
C _____
D _____

2X   6X/2 + 0.50   4X − 0.20   3X   2X − 0.3

A   B   C   D   0.70

7.0

© Cengage Learning 2013

20. In this CAD drawing, dimension **A** = 1.50″. Solve for dimension **X** to the nearest hundredth inch. (Hint: Write an expression for **X** in terms of **A**. Then evaluate each term using the given value of **A**.)

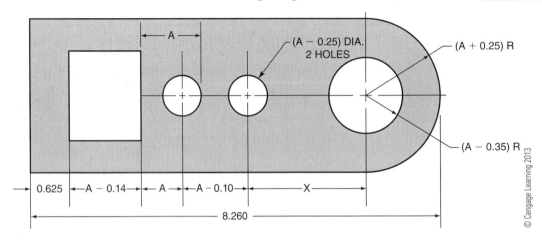

21. In this CAD illustration, dimension **B** is 1.35. What is dimension **Z**?

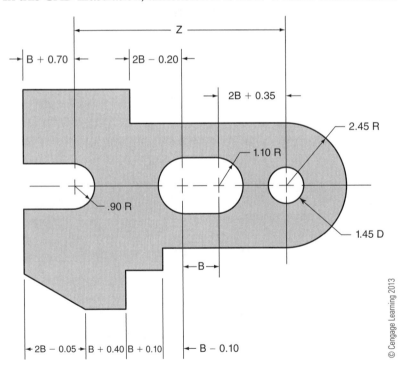

UNIT 30

Ratios

Basic Principles

Full-scale drawings are not always practical to create. Computer plots of large objects, therefore, must be reduced in order to be drawn on paper. And drawings of small objects must be enlarged to see the details clearly. Drawings are made to different scales, such as quarter-size, half-size, or double-size, and a fraction used to scale a drawing is called a ratio.

A *ratio* is a fraction that compares two numbers or similar quantities. Because ratios are fractions, they are usually expressed in simplest form, such as ⅔ or as the expression 2:1, which can be read as "2 to 1." A ratio such as 10:5 is not in simplest form and can be reduced to 2:1 by dividing both terms by 5. A ratio whose numerator (or denominator or both) is not a whole number, such as $\frac{2.5}{3}$, can be simplified by multiplying it by 1 in the form of the improper fraction ²⁄₂. The numerator and denominator of the product will then be whole numbers: $\frac{2.5}{3} \times \frac{2}{3} = \frac{5}{6}$.

NOTE: If a ratio involves measurements that have unlike units, such as feet and inches, first convert one of the two units to its equivalent value in terms of the second unit. Then, you can simplify the ratio as necessary, writing both numerator and denominator as whole numbers as described above.

EXAMPLE: Write the ratio 18 in. to 2 ft. in simplest whole number form.

STEP 1: Write the ratio as a fraction and include the units. The first term of the ratio is the numerator; the second term is the denominator.

$$18 \text{ in.} : 2 \text{ ft.} = \frac{18 \text{ in.}}{2 \text{ ft.}}$$

STEP 2: The measurement units in the ratio are not alike. Therefore, to express the ratio in simplest form, you need to either convert feet to inches (1) or inches to feet (2).

$$(1)\ \frac{18\text{ in.}}{2\text{ ft.}} = \frac{18\text{ in.}}{24\text{ in.}} \qquad \text{OR} \qquad (2)\ \frac{18\text{ in.}}{2\text{ ft.}} = \frac{1.5\text{ ft.}}{2\text{ ft.}}$$

STEP 3: Cancel out like units in each ratio and write each fraction in simplest (whole-number) form.

$$\frac{18\ \cancel{\text{in.}}}{24\ \cancel{\text{in.}}} = \frac{3}{4} \qquad \text{OR} \qquad \frac{1.5\ \cancel{\text{ft.}}}{2\ \cancel{\text{ft.}}} = \frac{1.5 \times 2}{2 \times 2} = \frac{3}{4}$$

The ratio 18 in. : 2 ft. $= \dfrac{3}{4}$.

Skill Problems

Express these ratios in simplest form. Be sure that the units in each ratio are alike before simplifying.

1. 3 cm to 20 mm _____

2. 6 in. to 2 ft. _____

3. 4 ft. to 6 in. _____

4. 2 yd. to 3 in. _____

Practical Problems

5. Gear A has 48 teeth, and gear B has 12 teeth. What is the ratio of the numbers of teeth between gear A and gear B? _____

6. In a drawing, a scale of 1″ = 1′-0″ is used. What is the ratio of the dimensions of the drawing to the dimensions of the actual object? _____

7. In a drawing, a scale of ¾″ = 1′-0″ is used. What is the ratio of the size of the drawing to the size of the actual object? _____

8. The scale ⅜″ = 1′-0″ is used in a drawing. What is the ratio of the lengths in the actual object to the lengths in the drawing? _____

9. If the scale ¼″ = 1″ is used, what is the ratio of the size of the actual object to the dimensions of the object being made? _____

10. It takes an experienced CAD drafter 16 hours to complete a set of working drawings. Another CAD drafter takes 48 hours to complete a similar set of working drawings. What is the ratio of the first drafter's time to the second drafter's time? _____

11. The gear ratio of the number of teeth on gear *A* to the number of teeth on gear *B* is 5:1. How many times faster does gear *B* turn than gear *A?*   _____

12. When the scale ½″ = 1′-0″ is used, what is the ratio of the size of the actual object to the dimensions in the drawing?   _____

13. Pulley *B* turns 9 times for every turn of pulley *A*. What is the ratio of diameter *A* to diameter *B?*   _____

14. An *inverse ratio* is the reciprocal of a given ratio. Gear *A* has 96 teeth, and gear *B* has 12 teeth. What is the inverse ratio between the number of teeth on gear *A* and the number of teeth on *B?*   _____

15. What is the ratio of the dimensions of a drawing to object if the scale ¼″ = 1′-0″ is used?   _____

16. CAD drafter *A* generates 24 architectural symbols in 12 hours. CAD operator *B* generates 24 symbols in 8 hours. What is the ratio of drafter *A*'s time to drafter *B*'s time?   _____

17. The scale 1½″ = 1′-0″ is used to draw an object. What is the ratio of the dimensions in the drawing to the dimensions of the actual object?   _____

18. If the scale 3″ = 1′-0″ is used to draw an object, what is the simplest whole number ratio between the size of the actual object and the dimensions of the drawing?   _____

CAD Problems

19. Measure dimensions **A**, **B**, and **C** on the CAD drawing below to the nearest 16th or 32nd inch. Use the scale 1:3 and calculate in inches the actual lengths of **A**, **B**, and **C**. Express the answers as mixed numbers in simplest form.

A _____

B _____

C _____

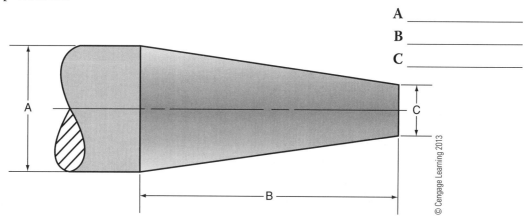

© Cengage Learning 2013

Practical Problems

20. Calculate the ratio of the circumference of the smaller circle to the circumference of the larger circle. Express the answer in simplest form. _____

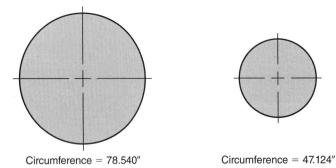

Circumference = 78.540"   Circumference = 47.124"   © Cengage Learning 2013

21. Measure dimensions **A, B, C,** and **D** on this CAD drawing to the nearest 8th or 16th inch. Use the scale 2:1 and calculate the actual lengths in inches of dimensions **A, B, C,** and **D** and express the answers as mixed numbers in simplest form.

A _____

B _____

C _____

D _____

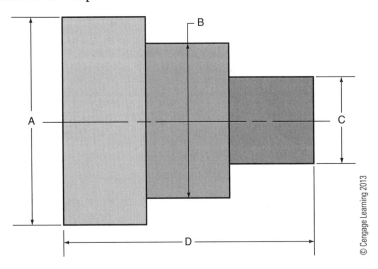

© Cengage Learning 2013

22. Measure the dimensions (sides and hole size) on the CAD drawing of the shim below. Using the scale 3:1, calculate the actual lengths in inches of dimensions A through H, expressed to the nearest hundredth.

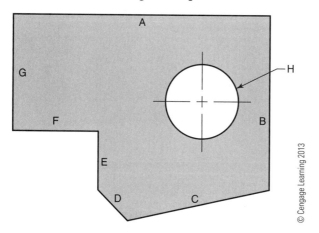

A _____

B _____

C _____

D _____

E _____

F _____

G _____

H _____

23. Find the simplest whole number ratio between the length of the diameter of hole **A** and the length of the diameter of hole **B**.

_____

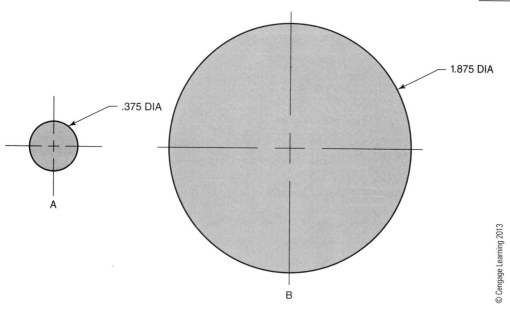

.375 DIA

1.875 DIA

UNIT 31

Proportions

Basic Principles

A proportion is a statement of equality between two ratios. There are two ways to write a proportion: either as an equation with a colon (:) between terms, as in $3 : 2 = 9 : 6$; or as two equal fractions, such as ½ = %. The terms in a proportion have special names: the *means* and the *extremes*. The second and third terms are the *means* (the *inner terms*). The first and fourth terms are the extremes (the *outer terms*).

$$3 : 2 = 9 : 6$$

Means

Extremes

Finding Cross-Products

RULE: In a valid proportion, the product of the means equals the product of the extremes.

When you write a proportion as two equal fractions and then multiply, the two products are called *cross-products*.

$$3 : 2 = 9 : 6$$

 Means

Extremes

$$3 \times 6 = 2 \times 9$$

Therefore, $\dfrac{3}{2} = \dfrac{9}{6}$ is a valid proportion because the cross-products are equal.

NOTE: The terms in a proportion can be inverted (flipped upside down),

$$3 : 2 = 9 : 6$$

$$\frac{3}{2} = \frac{9}{6} \Leftrightarrow \frac{2}{3} = \frac{6}{9}$$

or exchanged. $\frac{6}{2} = \frac{9}{3} \Leftrightarrow \frac{2}{6} = \frac{3}{9}$

In either case, if the cross-products are equal, the proportion is valid. And if the proportion is valid, the cross-products are equal.

Solving for a Missing Term

If you know three of the four terms in a proportion, you can calculate the fourth term. That's because a proportion is an equation, and you can use the **Properties of Equality** to solve for the missing term.

EXAMPLE: What is Z if $3 : 2 = 9 : Z$?

STEP 1:   Write the proportion as two equal fractions.

$$\frac{3}{2} = \frac{9}{Z} \quad \text{or} \quad \frac{2}{3} = \frac{Z}{9}$$

STEP 2:   Apply the cross-product rule.

$$2 \times 9 = 3 \times Z$$

$$18 = 3Z$$

STEP 3:   Divide both sides of the equation by 3.

$$6 = Z$$

So, the fourth term Z in this proportion is 6.

EXAMPLE: What is X if $\dfrac{X}{2} = \dfrac{9}{6}$?

STEP 1:   Apply the cross-product rule.

$$6 \times X = 2 \times 9$$

STEP 2:   Multiply the terms on each side of the equation.

$$6X = 18$$

STEP 3:   Divide both sides of the equation by 6.

$$X = 3$$

So, the fourth term X in this proportion is 3.

Skill Problems

For problems 1 to 4, solve for the missing term in each proportion.

1. $2:3 = 10:x$ _____

2. $A:5 = 36:20$ _____

3. $4:Y = 8:30.5$ _____

4. $4\frac{1}{2}:19\frac{1}{8} = ?:38\frac{1}{4}.$ _____

Practical Problems

The two triangles below are similar. This means that pairs of corresponding sides are proportional: for example, $c/b = A/B$. Use these figures to solve problems 5, 6, and 7.

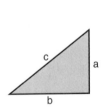
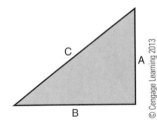

5.   Given that $B = 27$ mm, $b = 9$ mm, and $a = 5$ mm, what is A in millimeters?   _____

6.   Given that $a = 4$ inches, $c = 9$ inches, and $C = 42.75$ inches, what is A in inches?   _____

7.   In $\triangle ABC$, B is 145 centimeters, and C is 180 centimeters. In $\triangle abc$, b is 29 centimeters. What is c, in centimeters?   _____

© Cengage Learning 2013

8. A pinion gear with 15 teeth turns at 200 rpm (revolutions per minute). It is driven by a larger gear having 60 teeth. What is the rpm of the larger gear? (Hint: Write a proportion showing that the products of the number of teeth and the speed (rpm) of each gear are equal.)

9. A CAD operator realizes that it takes 19 minutes to plot a drawing. Working at the same rate, how long will it take the operator to plot eight hard copies of the same drawing?

10. A pump discharging 6 gallons of water per minute fills a tank in 30 hours. At this rate, how many hours does it take a pump discharging 20 gallons per minute to fill the tank?

11. In two weeks, five machinists assembled 12 machines. Working at the same rate, how many machinists were needed to assemble 60 machines in the same amount of time?

12. A gear with 12 teeth and turning at 1275 rpm is driving a gear with 60 teeth. What is the rpm of the larger gear?

13. A gear with 91 teeth and running at 240 rpm is being driven by a gear turning at 840 rpm. How many teeth does the driving gear have?

14. Two gears have a gear ratio of 2.8 : 1. If the larger gear has 98 teeth, how many teeth does the smaller gear have?

15. The diameter of a driven pulley is 9.25 inches and is rotating at 180 rpm. The driver pulley is rotating at 375 rpm. Determine the diameter of the driver pulley.

16. The diameter of pulley A is 9 inches, the diameter of pulley B is 4 inches, and the diameter of pulley C is 11.25 inches. The rpm of pulley A is 352 and the rpm of pulley D is 2376. What is the diameter of pulley D rounded to the nearest hundredth inch?

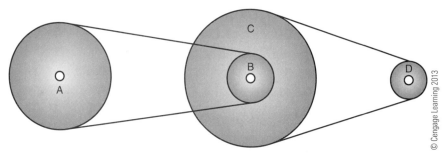

© Cengage Learning 2013

17. A CAD operator realizes that it takes her 40 minutes to plot a set of six drawings.

    a. Working at the same rate, how many sets of six drawings can she plot in 480 minutes?   a. _____

    b. How many drawings can she plot in 480 minutes?   b. _____

18. A structural drafter observes that a metal joint that is 8 feet long requires 40 rivets. How many rivets are required for a joint that is 5 feet long?   _____

19. If 10 CAD drafters can produce an average of 24 drawings in four days, how many CAD drafters, working at the same rate, will it take to average 36 of the same set of drawings in four days?   _____

20. In three weeks, six workers can assemble 32 products. Working at the same rate, how many workers are needed to assemble 544 products in the same amount of time?   _____

21. A CAD drafter observes that it takes 27 seconds to print seven drawings. At the same rate, how many minutes and seconds would it take to print 42 drawings?   _____

22. Assume that a freehand sketch can be made in $\frac{1}{10}$ of the time it takes to make an instrument drawing. If an instrument drawing takes $2\frac{1}{2}$ hours to complete, how many minutes should a freehand sketch take?   _____

23. If 1440 wedge blocks can be stamped out in 1 hour, determine how many blocks can be made at the same rate in 45 minutes and 15 seconds.   _____

UNIT 32

Formulas and Handbook Data

Basic Principles

A formula is a mathematical statement of equality between variable expressions. Most mathematical formulas relate to particular properties of figures, such as the perimeters and areas of geometric figures, and values of particular constants, such as π. When stating a formula, you need to define the meanings of all of the symbols and variables in the formula.

Handbook data are found within the five tables in Section II of the Appendix on page 392 of this book. These tables include lists of equivalent units of measures, equivalent fractions and decimals, common mathematical formulas, and other important information.

As a drafter, you will often need to use formulas and data when making drawings. For example, formulas are used to obtain dimensions for layout purposes or for the manufacture of a product and the construction of a building. Handbooks and technical publications contain technical information that must be referenced to determine dimensions, sizes, costs, weights, and so on.

EXAMPLES:

- The formula for calculating the approximate value of π is $\pi = \dfrac{C}{D}$, where C is the circumference of a circle, and D is the diameter of the circle.

- A formula for calculating the area A of a triangle is $A = \frac{1}{2}bh$, where b is the length of a base (side), and h is the altitude (height) to that base.

Skill Problems

Use TABLE I, EQUIVALENT U.S. CUSTOMARY AND METRIC UNITS OF MEASURE, in Section II, page 392 of the Appendix to answer questions 1 to 5.

1. What is the millimeter equivalent of 7 inches? _____

2. What is the inch equivalent of 4 millimeters? _____

3. What is the millimeter equivalent of 3 inches? _____

4. What is the area equivalent of 6 square inches in square centimeters? _____

5. What is the area equivalent of 4 square centimeters to the nearest hundredth square inch? _____

Use TABLE IV, POWERS AND ROOTS OF NUMBERS (1 TO 100), on page 394 in Section II of the Appendix to answer questions 6 to 10.

6. What is the circumference, to the nearest hundredth inch, of a circle whose diameter is 11 inches? _____

7. What is the diameter, to the nearest hundredth inch, of a circle whose circumference is 23.78 inches? _____

8. What is the diameter, to the nearest tenth centimeter, of a circle whose area is 18.1 cm²? _____

9. What is the circumference in millimeters to the nearest hundredth, of a circle whose diameter is 7 millimeters? _____

10. What is the radius in millimeters of a circle whose area is 5281 mm²? _____

Practical Problems

11. The formula for calculating the minimum depth **D** when tapping a thread in cast iron is 1½ **D**. What is the minimum depth in inches for tapping a ⅝ − 16 NC thread in cast iron? Express the answer in simplest form. _____

12. Calculate to the nearest hundredth inch, the diameter **D** of a circle if the area **A** of the circle is 628.32 sq. in., and $D \approx \sqrt{\dfrac{A}{0.7854}}$. _____

13. When making a full development of the slope of a right circular cone the drafter lays out the measure of angle X using the formula,
$X° = \dfrac{R}{S} \times 360°$. Determine the measure of angle X if
dimension $\mathbf{R} = 60$ mm and dimension $\mathbf{S} = 135$ mm.

_____

14. Heron's formula for finding A, the area of a triangle, is
$A = \sqrt{s(s - a)(s - b)(s - c)}$, where s is the semi-perimeter
of the triangle, $s = \dfrac{a + b + c}{2}$, and a, b, and c are the lengths of the
sides of the triangle.

a. What is the semi-perimeter of a triangle whose
sides are 5, 12, and 13?   a. _____

b. What is the area of the triangle to the nearest whole number?   b. _____

c. What kind of triangle is this triangle?   c. _____

15. You can use the formula $A = 8sh$ to determine a stretchout (development) of a right octagonal prism, where s is the side dimension, and h is the height of the prism. Calculate to the nearest square inch, the area A of a stretchout to the nearest hundredth if $s = 2.125$ inches, and $h = 4.45$ inches.

_____

CAD Problems

16. The figure below is a regular pentagon. Use the formulas next to the pentagon and calculate the number of degrees in angles A, C, and E.

N = number of sides

$\angle C = \dfrac{360°}{N}$

$\angle A = \dfrac{N - 2}{N} \times 180°$

$\angle A$ _____
$\angle C$ _____
$\angle E$ _____

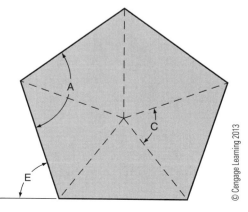

© Cengage Learning 2013

17. Drafters follow standard rules and a definite order in almost all work they do. If exact measurements are not known, approximate sizes and formulas are used. For example, this CAD drawing of a ¾"-diameter hexagonal bolt and nut was drawn using known formulas. Dimensions **B**, **A**, and **C** are based on the diameter **D** of the fastener.

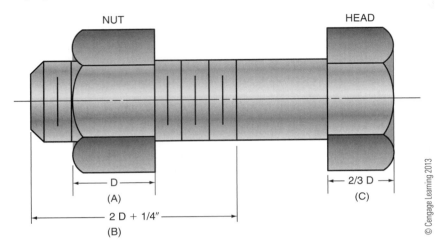

© Cengage Learning 2013

a. What is the threaded length of bolt **B**?

b. What is the thickness of nut **A**?

c. What is the thickness of bolt head **C**?

a. _____

b. _____

c. _____

18. A formula for calculating the equal lengths of d, the diagonals of a square, is $d = \sqrt{2s^2}$, where s is the length of each side of the square. Calculate the length of the diagonals in the square below to the nearest hundredth.

_____

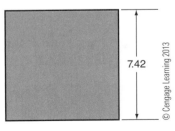

7.42

© Cengage Learning 2013

19. Calculate, in inches, the circumference of the CAD drawing below of a 6.75-inch diameter disk cam. The circumference of the cam equals the stretchout length of this displacement diagram. Use 3.1416 as the value of π and round the answer to the nearest thousandth. _____

DISPLACEMENT DIAGRAM

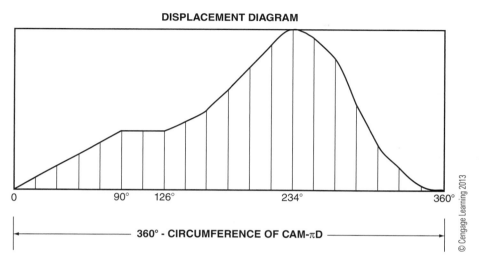

20. Use the formula $T = \dfrac{D - d}{L}$ and calculate to four decimal places the amount of taper per inch, T, for this CAD view of a tapered shaft, where D and d represent diameters of the shaft, and L represents its length. _____

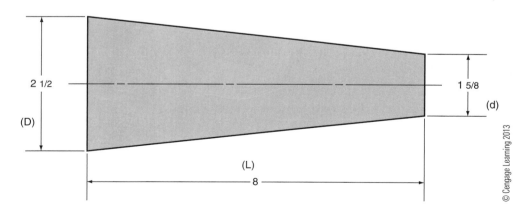

21. A drafter must determine several measurements before a gear can
be drawn. Most gear-cutting data can be found if the diametral pitch
P and the number of teeth **N** in the gear are known. In this gear, the
diametral pitch **P** is 4, and the number of teeth **N** is 24. Use each given
formula to solve parts *a* through *e* that follow.

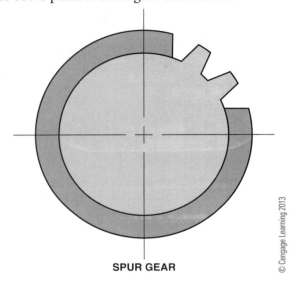

SPUR GEAR

© Cengage Learning 2013

a. Calculate the pitch diameter **D** to the nearest whole number.

$$D = \frac{N}{P}$$

a. _____

b. Calculate the outside diameter **O** of the gear to the nearest tenth.

$$O = \frac{N + 2}{P}$$

b. _____

c. Calculate the addendum **A** to the nearest hundredth (The *addendum* is the radial distance between the pitch circle to the top of a
gear tooth.).

$$A = \frac{1.000}{P}$$

c. _____

d. Calculate the thickness of the gear tooth **T** to the nearest thousandth.

$$T = \frac{1.5708}{P}$$

d. _____

e. Calculate the whole depth of the gear tooth **W** to the nearest thousandth.

$$W = \frac{2.157}{P}$$

e. _____

UNIT 33

Averages

Basic Principles

An average is a statistical measure, and a very common type of average is called the *mean*. This number is a statistical measure of the "middle" of a set of data. You can calculate the mean of a set of numerical data by using the formula below.

$\overline{X} = \dfrac{x_1 + x_2 + x_3 + \dots + x_n}{n}$, where $\overline{X}$ represents the mean, n is the number of data values, and $x_1, x_2, \dots x_n$ are the values in the data set.

EXAMPLE: What is the average of 46, 17, 21, and 16?

In this example, $n = 4$, and the data are 46, 17, 21, and 16. Substitute these values in the equation and solve for the mean.

$$\overline{X} = \frac{46 + 17 + 21 + 16}{4}$$
$$= \frac{100}{4}$$
$$= 25$$

The mean for these data is 25, the number in the "middle."

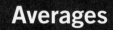 46 ⊞ 17 ⊞ 21 ⊞ 16 ⊟ ⊡ 4 ⊟

NOTE: If your calculator has a statistical mode, you can enter the expression shown below. The $\sigma\Sigma$ stands for a sum. But first, in STAT mode, press the 2nd and CSR keys to clear any data that may still be in the memory of the calculator.

46 Σ+ 17 Σ+ 21 Σ+ 16 Σ+ 2nd X̄

323

Practical Problems

1. Three CAD drafters receive hourly wages of $26.50, $23.25, and $29.75, respectively. What is the average hourly pay for these drafters? _____

2. A CAD drafter assigns the following number of levels to certain phases of an engineering project: 30, 12, 26, 42, 37, 22, and 20. Calculate the average number of levels that she assigns to this project. _____

3. Over a 5-day period, a drafter logged the following number of work-related miles: 48, 33, 28, 40, and 37. Calculate to the nearest hundredth, the average number of miles he traveled per day. _____

4. A mechanical drafter completed the following number of drawings over a period of four weeks: 19, 17, 16, and 15. What was the average number of drawings he completed each week? _____

5. A class of drafting students received the following scores on their first exam: 78, 98, 82, 66, 92, 83, 79, 87, and 82. What was the average score on this exam for these students? _____

6. An inexperienced CAD drafter earns $12.50 per hour. During a 5-day period, she works the following numbers of hours: 8, $5\frac{1}{2}$, 9, $6\frac{1}{4}$, and $7\frac{3}{4}$. Calculate her average daily earnings. _____

7. Calculate the average of $\frac{2}{3}$, $\frac{5}{16}$, $\frac{7}{12}$, and $\frac{3}{4}$. Express the answer as a proper fraction in simplest form. _____

8. A detail drafter completed 96 drawings during a 1-year period. Calculate the average number of drawings the drafter completed each month. _____

9. An architect designs a building with five different types of doors. The doors cost $185.50, $343.75, $1,255.30, $737.45, and $279.65 respectively. Calculate the average cost of all of the doors. _____

10. On a drafting test, students achieved the following scores: two scored 87%; three scored 76%; four scored 93%; five scored 81%; and one scored 55%. To the nearest hundredth, what was the average test score for all of the tests?

    _____

11. A CAD drafter completes 11 drawings on Monday, 9 on Tuesday, 13 on Wednesday, and 17 on Thursday. How many drawings must be completed on Friday to average 13 drawings each day?

    _____

12. A drafter spent the following amounts for equipment during three months: $16.20, $35.78, $78.36, $42.83, and $56.52. What is the average amount of money, to the nearest cent, that he spent?

    _____

13. Over a period of seven years, a drafter missed the following number of working days each year: 16, 7, 11, 9, 7, 10, and 8. What was the average number of working days he missed per year?

    _____

14. What is the average weight in pounds to the nearest hundredth, of six bars of metal that weigh: 6½ lb., 3¼ lb., 2⅞ lb., 7 lb., 5¾ lb., and 12⅜ lb.? _____

15. A CAD drafter totals the number of symbols from the symbol library he used on seven drawings. The total number of symbols he used for each drawing were 17, 61, 32, 43, 29, 56, and 35. What was the average number of symbols he used on all these drawings?

CAD Problems

16. What is the average length of the lines below?

    _____

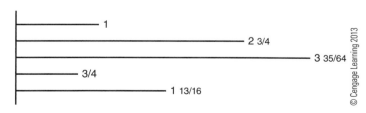

1

2 3/4

3 35/64

3/4

1 13/16

17. Calculate the average height and width of the CAD drawing of the step below. Then calculate the averages of the diameters of all the holes in the drawing. Round each answer to the nearest thousandth.

Average height _____

Average width _____

Average diameter _____

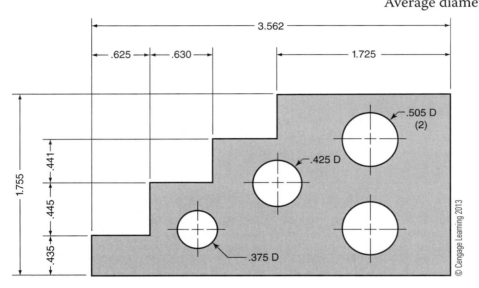

18. What is the average length of the dimensions on this CAD drawing of a template? Round the answer to the nearest hundredth.

_____

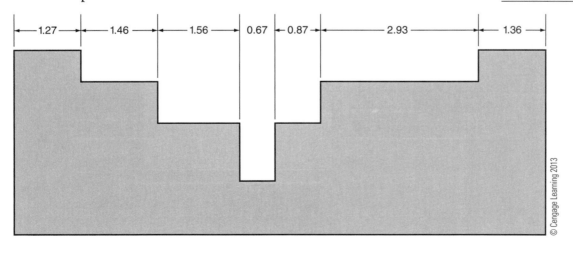

SECTION

Graphing

UNIT 34

Basic Principles

Drafters often display drawings on coordinate grids that are part of a *rectangular coordinate system*. This system is a way to locate points in a plane using pairs of numbers. Two perpendicular lines, called *axes*, divide the plane into four regions, called *quadrants*. The horizontal and vertical axes intersect at a point called the *origin*. (The horizontal axis is often referred to as the *x*-axis, and the vertical axis is often referred to as the *y*-axis.) In drafting, because all measurements are positive, points in the coordinate plane are represented along one or both axes and within the first quadrant.

The two axes in a rectangular coordinate system usually have the same scale starting at (0, 0), the origin. This results in the plane being divided into congruent rectangles. Hence the name, *rectangular coordinate system*.

You can specify the location of any point in the plane using an *ordered pair* of numbers called *coordinates*. In drafting, the coordinates of an ordered pair can be written with or without parentheses: for example, **(0, 0)** or **0, 0** both indicate the coordinates of the origin.

Coordinates of points in quadrant I and on either or both positive axes have the following properties.

- The first number in an ordered pair represents the *horizontal* distance of the point to the right of the vertical axis or the distance along the *y*-axis when $x = 0$.
- The second number in an ordered pair represents the *vertical* distance of the point above the horizontal axis or the distance along the *x*-axis when $y = 0$.

Using the location of the origin, the coordinates of a given point and the scales along the axes, you can determine the lengths of segments and scale the dimensions of objects as necessary.

EXAMPLE: Use the drawing of the object on the grid below to identify the two missing coordinates of the vertices of the rectangle and the coordinates of the vertices of the quadrilateral inside the rectangle.

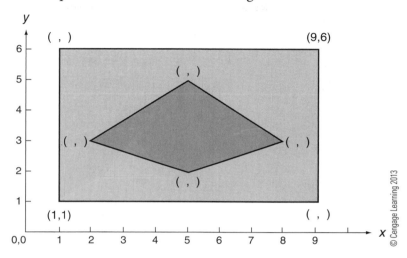

The scales on each axis are equal, and each interval equals 1 unit. Line up each vertex with the scales along each axis.

- The coordinates of the upper left vertex of the rectangle are (1, 6).
- The coordinates of the lower right vertex of the rectangle are (9, 1).
- The coordinates of the vertices of the quadrilateral inside the rectangle are (2, 3), (5, 5), (8, 3), and (5, 2).

EXAMPLE: Use the figure above and calculate the length and width of the rectangle on the grid if each interval between grid marks represents 0.5 units.

	Original dimension (units)	Scaled dimension (units)
Length	9 − 1 = 8	0.5 × 8 = **4.0**
Width	6 − 1 = 5	0.5 × 5 = **2.5**

CAD Problems

Use the drawing of the object on the grid that follows to solve problems 1 through 3.

NOTE: The numbers on the figure mark 20 vertices moving clockwise from the ordered pair labeled **START PT. 2, 1**. (The origin is the circled mark showing **0, 0** in the lower left corner.)

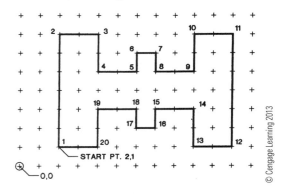

1. Each space between grid marks on this coordinate grid represents 3 units. What is the perimeter of the figure expressed in units? _____

2. Suppose that the grid spacing on the drawing above equals ⁷⁄₁₆″. What is the perimeter in inches of the figure? Express the answer as a mixed number in simplest form. _____

3. Suppose that each interval between the grid marks represents 8 inches. If the drawing of the figure is to be scaled down by a factor of 3, what is the resulting height and width of the figure in inches?   Height _____
     Width _____

Use the drawing of the object below to solve problems 4 to 6. The numbers 1 through 20 on the object represent the location of the 20 vertices of the figure starting at (2, 1).

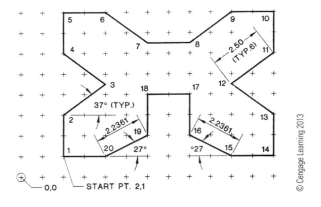

4. If the actual object has a width of 8¼″ and a height of 6⅛″, what is the
    spacing in inches between each grid mark?   _____

5. Suppose that the size of the object is to be scaled down by a factor of
    4. If the grid spacing in the drawing is 1.0″, calculate to the nearest
    hundredth inch the new height and width of the object.   Height _____

                                                                  Width _____

6. Suppose that the grid spacing on the drawing is 0.875″. What is the
    perimeter of the figure in inches to four decimal places?   _____

7. The figure drawn on the coordinate plane below is to be scaled up by a
    factor of 7. Calculate the new height and width of the polygon and the
    new diameter of the circle.   Height _____

                                                                  Width _____

                                                                  Diameter _____

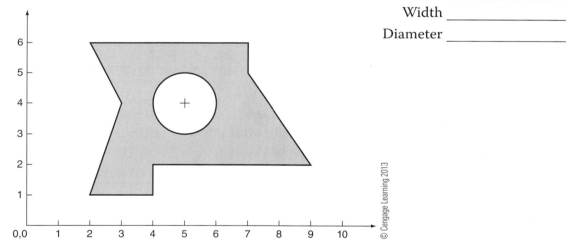

© Cengage Learning 2013

8. This CAD drawing shows the front view of object ***ABCDE*** plotted on a rectangular plane. If the size of the object is to be scaled up by a factor of 5, what will be the new height and width of the object?   Height _____

Width _____

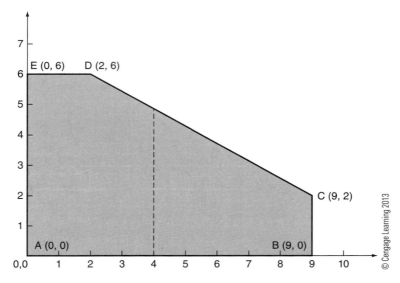

9. The unit spacing on each axis below represents ⁷⁄₁₆″. Using this scale, what is the height and width of the object in inches? Express the answers as mixed numbers in simplest form.   Height _____

Width _____

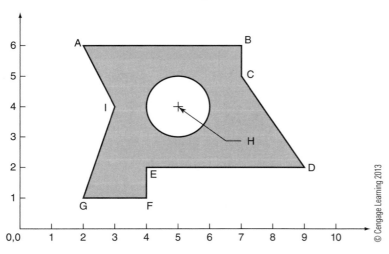

10. Each interval on the axes below represents ¾". If the figure is to be scaled up by a factor of 4, what will be its new height and width in inches?

Height _____

Width _____

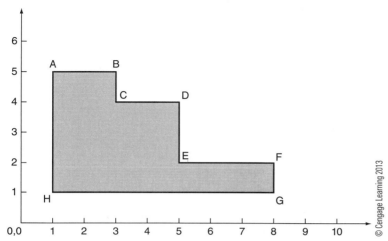

11. The figure in the drawing below is to be inserted into a drawing that is seven times larger than its existing size. If the current grid spacing is 0.375, find the new height and width of the figure after it is inserted. Round the answers to the nearest hundredth.

Height _____

Width _____

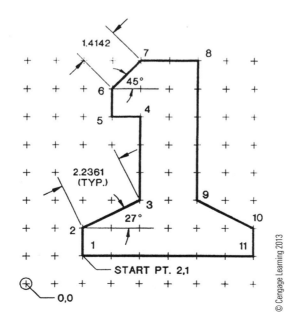

12. The numbers 1 through 20 on this drawing represent the location of the vertices of the figure. Determine the area of the figure, in square inches, if the grid spacing is 0.375". Round the answer to three decimal places.

_____

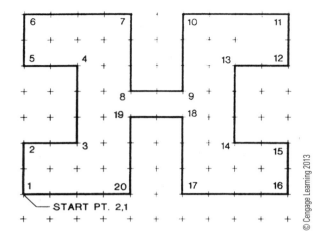

UNIT 35

Statistical Graphs

Basic Principles

A statistical graph is a way to visually describe, compare, and analyze data. The most common types of statistical graphs are bar graphs, line graphs, and circle graphs (pie charts). The type of graph used to represent data from one or more variables depends on the nature of the data. Bar graphs and line graphs are most often used to compare relationships between sets of data, such as length and width. Circle graphs are used to display relationships among parts of a whole, such as the number or percentage of drafters working in one department.

NOTE: To create a circle graph, you need to calculate the number of degrees in the central angle of each part of the graph. You can do this by multiplying 360° by each given percent).

Practical Problems

1. This circle graph represents the total expenses of a company. Use the graph to answer the following questions.

Company Expenses

© Cengage Learning 2013

a. Which labeled part of the circle graph represents the greatest
   expense?   **a.** _____

b. Which labeled part represents the least expense?   **b.** _____

c. What is the percent difference between the greatest expense and
   least expense?   **c.** _____

2. This line graph displays the numbers of appliance shipments in
   thousands of units made during a 12-month period. Use the graph to
   answer the following questions.

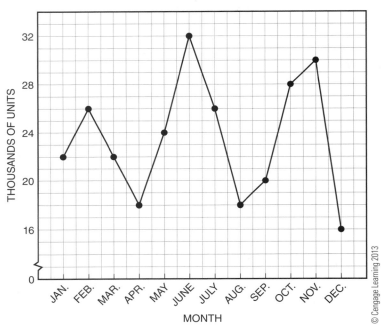

a. During which two months did the shipments decrease the most?   **a.** _____

b. During which month did the greatest number of shipments take
   place?   **b.** _____

c. During which month did the fewest number of shipments take
   place?   **c.** _____

    d. During which month was the shipment halfway between the
       lowest and greatest numbers of shipments?   **d.** _____

    e. During which month were 20,000 units shipped?   **e.** _____

3. This bar graph shows the number of jobs completed during a 6-year
    period. Use the graph to answer the following questions.

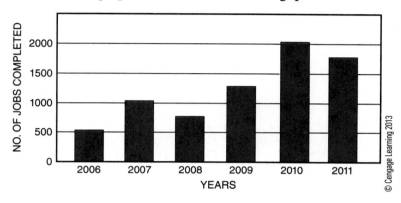

© Cengage Learning 2013

    a. In which year were the greatest number of jobs completed?   **a.** _____

    b. In which year were the fewest number of jobs completed?   **b.** _____

    c. In which year was the number of jobs completed halfway between
       the years with the greatest and fewest numbers of completed
       jobs?   **c.** _____

    d. What was the difference between the number of jobs completed
       for the years 2006 and 2011?   **d.** _____

4. At one point in time, the raw steel production in the U.S. by types was open hearth—55%; electric furnace—17%; and basic oxygen—28%. Use the figure below, a protractor, and a ruler to create a circle graph that best represents these data. Round degree measures of each part to the nearest tenth as necessary and label the part of the circle that corresponds to each type of production. (Hint: Calculate the percent of 360 degrees represented by each type of production.).

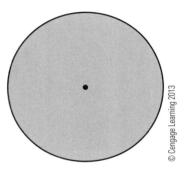

5. The line graph below displays the number of hours per month that a company spent on a special drafting project one year. Use the graph to answer the following questions.

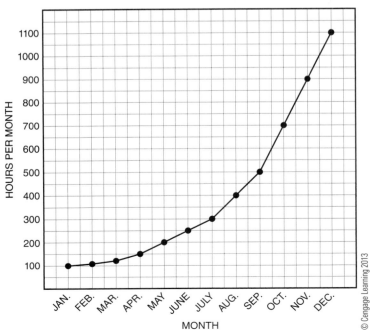

ANNUAL REPORT OF HOURS SPENT
ON SPECIAL DRAFTING PROJECT

a. How many hours were spent on the project during November?   **a.** _____

b. In which month were the hours spent seven times the number of hours spent in January?   **b.** _____

c. In what month were 150 hours spent on the project?   **c.** _____

d. Which three months had the largest increase in the number of hours worked?   **d.** _____

6. Eight types of jobs make up a given engineering department. Create a bar graph on the grid below to show the distribution of workers in each type of job in the department.

Drafters	26%	Engineers	14%
Checkers	8%	Admin. Asst.	6%
Technicians	8%	Clerks	4%
Architects	10%	CAD operators	24%

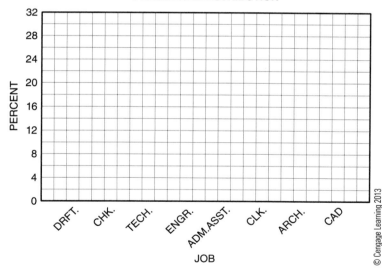

PERSONNEL DISTRIBUTION

7. A summer audit of the Acme Wholesale Company revealed its net gain in production during six months. On the grid below, create a line graph that displays the data collected during this 6-month period: in April, the net gain was $15,000; May—$20,000; June—$10,000; July—$25,000; August—$30,000; and September—$27,500.

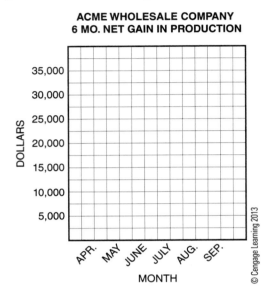

ACME WHOLESALE COMPANY
6 MO. NET GAIN IN PRODUCTION

© Cengage Learning 2013

8. Use the figure below, a protractor, and a ruler to create a circle graph that displays the five types of drawings completed by members of a drafting department. Round the angle measures in each part of the graph to the nearest whole number of degrees. (Hint: Calculate the approximate percentage of 360 degrees each type of drawing represents.) Label what each part of the graph represents.

Machine drawings	45%	Foundry drawings	19%
Structural drawings	14%	Sheet metal drawings	15%
Pattern drawings	7%		

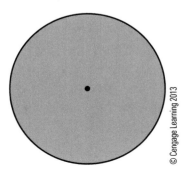

© Cengage Learning 2013

SECTION

Applied Trigonometry

UNIT 36

Right Triangles

Basic Principles

Trigonometry is the branch of mathematics that deals with triangles and the measures of their sides and angles. Trigonometry is very important to the work of all drafters. When dealing with triangles, drafters and CAD operators must understand the mathematics of trigonometric operations and their relationships.

A right triangle, as the name implies, is a triangle that contains one right angle. Its measure is 90°. Because the sum of the measures of the three angles of a triangle is 180°, the other two angles in a right triangle are acute. Each one has a measure less than 90 degrees, and the sum of their measures is 90°. This means that these two angles are complementary.

The sides in a right triangle have special names, based on where they lie with respect to the angles of the triangle. Study the information below carefully to understand how the parts of a right triangle relate to one another.

- The side opposite the right angle is called the *hypotenuse.*
- The other two sides are called *legs.*

The legs in a right triangle lie *opposite* and *adjacent* to each of the acute angles of the triangle.

- The leg *opposite* an acute angle is not one of the sides of the angle.
- The leg *adjacent* to an acute angle is one of the sides of the angle.

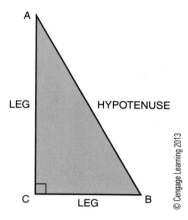

When solving trigonometry problems, it is very important to identify the relationships among the angles and sides of a right triangle. These relationships form the bases of the definitions of the trignometric ratios listed below.

The Trigonometric Ratios

The six trigonometric ratios and their abbreviations are the sine (sin), the cosine (cos), the tangent (tan), the cotangent (cot), the secant (sec), and the cosecant (csc). Most common trigonometry problems involve the sin, cos, and tan ratios.

Here are the definitions of the six trigonometric ratios for $\angle A$ and $\angle B$ in right triangle $\triangle ABC$, with right angle at C and hypotenuse c.

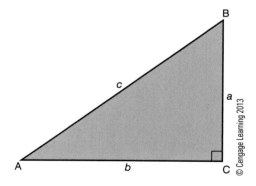

- The sine (sin) of an acute angle is the ratio between the leg opposite the angle and the hypotenuse.

$$\sin A = \frac{\text{opposite leg}}{\text{hypotenuse}} = \frac{a}{c} \qquad \sin B = \frac{\text{opposite leg}}{\text{hypotenuse}} = \frac{b}{c}$$

- The cosine (cos) of an acute angle is the ratio between the leg adjacent to the angle and the hypotenuse.

$$\cos A = \frac{\text{adjacent leg}}{\text{hypotenuse}} = \frac{b}{c} \qquad \cos B = \frac{\text{adjacent leg}}{\text{hypotenuse}} = \frac{a}{c}$$

- The tangent (tan) of an acute angle is the ratio between the leg opposite the angle and the leg adjacent to the angle.

$$\tan A = \frac{\text{opposite leg}}{\text{adjacent leg}} = \frac{a}{b} \qquad \tan B = \frac{\text{opposite leg}}{\text{adjacent leg}} = \frac{b}{a}$$

- The cotangent (cot) of an acute angle is the ratio between the leg adjacent to the angle and the leg opposite the angle.

$$\cot A = \frac{\text{adjacent leg}}{\text{opposite leg}} = \frac{b}{a} \qquad \cot B = \frac{\text{adjacent leg}}{\text{opposite leg}} = \frac{b}{a}$$

- The secant (sec) and cosecant (csc) of an acute angle are reciprocals of the cosine and the sine of the angle respectively.

$$\sec A = \frac{\text{hypotenuse}}{\text{adjacent leg}} = \frac{c}{b} \qquad \csc A = \frac{\text{hypotenuse}}{\text{opposite leg}} = \frac{c}{a}$$

$$\sec B = \frac{\text{hypotenuse}}{\text{adjacent leg}} = \frac{c}{a} \qquad \csc B = \frac{\text{hypotenuse}}{\text{opposite leg}} = \frac{c}{b}$$

Finding the Trig Ratio of an Angle

The trig ratio of an acute angle can be any real number, depending on the ratio. Nearly all trigonometric values are decimal approximations that you can round as necessary. You can find the sine, cosine, and tangent ratios of any acute angle using a trigonometry table or a calculator.

EXAMPLE: What is sin 38°? Express the answer to four decimal places.

 38 [SIN]

sin 38° ≈ 0.6157

EXAMPLE: What is cos 42°? Express the answer to four decimal places.

 42 [COS]

cos 42° ≈ 0.7431

EXAMPLE: What is sec 42°? Express the answer to four decimal places.

 42 [COS] [⅟ₓ]

NOTE: To find the reciprocal of a trig ratio use the [⅟ₓ] key on a calculator.

Finding the Angle Measure Given a Trig Ratio

If you know the trig ratio for a given angle, you can use a trig table or a calculator to find the measure of the angle. This usually requires finding the decimal value for that ratio in the body of a trig table and reading across to the angle measure, or using the [2nd] function key on a calculator.

EXAMPLE: If sin A = 0.876, what is the measure of $\angle A$? Round the answer to the nearest whole number.

 .876 [2nd] [SIN]

$\angle A \approx 61°$

EXAMPLE: If tan B = 1, what is the measure of $\angle B$?

1 [2nd] [TAN]

$\angle B = 45°$

Cofunctions

The two acute angles in an right triangle are *complementary*. This means that the sum of their measures is 90°. Further, if angles A and B are complementary, then $\angle A$ is the complement of $\angle B$, and $\angle B$ is the complement of $\angle A$.

The *cofunction* of an angle refers to the trigonometric ratio of the complement of the angle. The following expressions show the relationships between the cofunctions of complementary angles A and B in right triangle ABC, where $\angle C$ is the right angle.

$$\sin A = \frac{a}{c} = \cos B \qquad \sin B = \frac{b}{c} = \cos A$$

$$\tan A = \frac{a}{b} = \cot B \qquad \tan B = \frac{b}{a} = \cot A$$

$$\sec A = \frac{c}{b} = \csc B \qquad \sec B = \frac{c}{a} = \csc A$$

EXAMPLE: Given $\sin 30° = 0.5000$, what is $\cos 60°$?

The two angles are complementary because $30° + 60° = 90°$. Therefore, $\cos 60°$ is the cofunction of $\sin 30°$.

So, $\cos 60° = 0.5000 = \sin 30°$.

EXAMPLE: Triangle *ABC* is a right triangle with right angle at *B*. The measures of its sides are 6, 8, and 10. Use a calculator to determine the measures of angles *A* and *C* to the nearest degree.

NOTE: The right angle in a right triangle is not always labeled *C*.

STEP 1:   Use the definition of the sine ratio and write sin *C* and sin *A* using the lengths of their opposite legs and the hypotenuse, the longest side in the triangle.

$$\sin C = \frac{8}{10} = 0.8000 \qquad \sin A = \frac{6}{10} = 0.6000$$

STEP 2:   Use a trig table or a calculator to determine the measures of angles *A* and *C*. Round the answers to the nearest whole number of degrees.

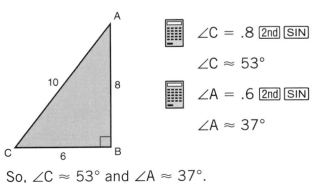

∠C = .8 [2nd] [SIN]

∠C ≈ 53°

∠A = .6 [2nd] [SIN]

∠A ≈ 37°

So, ∠C ≈ 53° and ∠A ≈ 37°.

EXAMPLE: Determine the length of side X in right triangle *ABC*.

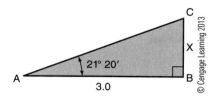

STEP 1:   Side X is the leg opposite ∠*A,* and $\overline{AB}$ is the leg adjacent to ∠*A*. So, use tan *A* to determine X.

$$\tan A = \frac{X}{AB}$$

$$\tan (21°20') = \frac{X}{3}$$

STEP 2:   Cross multiply the terms in the proportion.

$$\frac{\tan (21°20')}{1} = \frac{X}{3}$$

$$1X = 3 \times \tan (21°20')$$

To evaluate X, you need to determine the tangent of 21° 20′ using either a calculator or a trig table. Each method is described below.

STEP 3:   **Method 1: Using a Calculator**

You can work with the measure of the angle in degrees and minutes or with the measure expressed as a decimal in degrees only. Each is explained in parts A and B that follow.

A. Convert 20′ to degrees.

$$\frac{20'}{60'} = \frac{1}{3} \approx 0.33°$$

So, 21° 20′ ≈ 21.33°.

Now determine tan 21.33°.

  21.33 [TAN]

B. You can also use the ⌊2nd⌋ and ⌊DMS ▸ DD⌋ function keys to convert an angle measured in degrees, minutes, and seconds (DMS) into decimal degrees (DD). Enter 21°20′ as 21.20. Then determine the tangent of the angle.

 21.20 ⌊2nd⌋ ⌊DMS ▸ DD⌋ ⌊TAN⌋

In both cases, tan 21° 20′ ≈ tan 21.33° ≈ 0.3905.

Method 2: Using a Trig Table

Again, you can work with the measure in degrees only (part A) or in degrees and minutes (part B).

A. If the degree measure of an angle is a decimal, such as 21.33°, you can *interpolate* to approximate the value of the trig ratio. To interpolate, you need to write a proportion using the differences between angle measures and decimal values. In the example below, tan 21.33° on the left lies between tan 21° and tan 22°. The ratio can be written as 0.33:1.00. The ratio on the right is y: 0.0201 where y is a number between 0 and 0.0201.

$$
1.00 \left[\begin{array}{l} .33 \left[\begin{array}{l} \text{tan } 21° = 0.3839 \\ \text{tan } 21.33° = y \end{array} \right. y \\ \text{tan } 22° = 0.4040 \end{array} \right] .0201
$$

Cross multiply to solve for y.

$$
\frac{0.33}{1.00} \approx \frac{y}{0.0201}
$$

$$
1y \approx 0.33 \times 0.0201 \approx 0.0066
$$

The tangent of 21.33° increases between tan 21° and tan 22°, so add the value of y to the lower value 0.3839 to approximate tan 21.33°.

So, tan 21° 20′ ≈ tan 21.33°

$$
\approx 0.3839 + 0.0066
$$

$$
\approx 0.3905
$$

B. If an angle measure is in degrees and minutes, such as 21° 20′, you can still interpolate. For this problem, the ratio of the differences on the left is in minutes: 20′ : 60′. The ratio of the differences on the right is still y : 0.0201 as before.

$\tan 21°0′ \approx 0.3839$
$\tan 21°20′ = y$
$\tan 22°0′ \approx 0.4040$

Cross multiply to determine y.

$$\frac{20}{60} \approx \frac{y}{0.0201}$$

$y \approx 0.0067$

Add the value of y to the lower value, 0.3839, to approximate $\tan 21° 20′$.

$0.3839 + 0.0067 \approx 0.3905$

STEP 4:   Multiply 3 and 0.3905 to solve for side X in $\triangle ABC$. Round the answer to the nearest tenth.

$$\frac{X}{3} = \tan 21°20′$$

$X \approx 3 \times 0.3905 \approx 1.2$

So, side X in $\triangle ABC$ is about 1.2.

Practical Problems

1. Determine the value of sin 76° to five decimal places.   _____

2. Determine the value of tan 26° to five decimal places.   _____

3. Determine the value of cos 55° to five decimal places.   _____

4. Determine to the nearest degree the measure of the angle whose sine equals 0.94552.   _____

5. Determine to the nearest degree the measure of the angle whose tangent equals 1.4281.   _____

6. Determine to the nearest degree the measure of the angle whose cosine equals 0.97029.   _____

7. What is the cofunction of sin 29°? _____

8. What is the cofunction of cot 21°? _____

9. What is the cofunction of cos 54°? _____

10. Use interpolation to determine the sine of 27° 13′. Round the answer to five places. _____

11. Use interpolation to determine the cosine of 27° 13′. Round the answer to five decimal places. _____

12. To the nearest degree what is the measure of the angle whose cosine equals 0.76604? _____

13. To the nearest degree what is the measure of the angle whose tangent equals 4.4494? _____

14. What is the length in inches of the base of a right triangle with a 35° base angle and a 12-inch altitude to that base? Express the answer to the nearest hundredth. (Hint: Draw and label the triangle showing the given parts.) _____

15. What is the length in inches of the hypotenuse of a right triangle that has a 60° base angle and an 11-inch base? _____

CAD Problems

16. Calculate the measure of angle *X* in this right triangle. _____

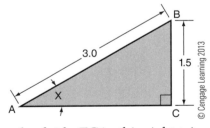

17. Calculate the length of side *FG* in this right triangle. _____

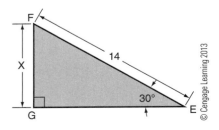

18. Calculate dimension X of this shelf brace to the nearest tenth. _____

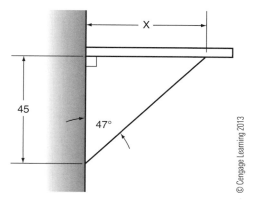

19. Calculate distance **AC** between the center of hole **A** to the center of hole **C** and distance **BC**, between the center of hole **B** to the center of hole **C**. Express each answer to three decimal places.

AC _____

BC _____

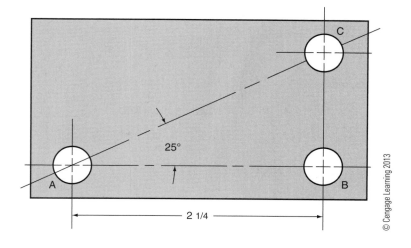

20. Determine the measure of ∠X needed to machine this shaft as illustrated. Express the answer to the nearest minute, using interpolation as necessary. _____

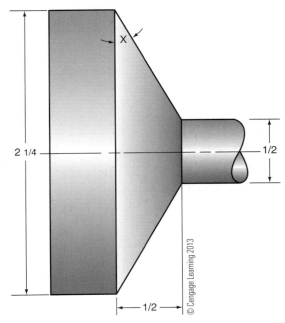

2 1/4

1/2

1/2

© Cengage Learning 2013

21. Calculate the center distance X on this gasket to the nearest hundredth. _____

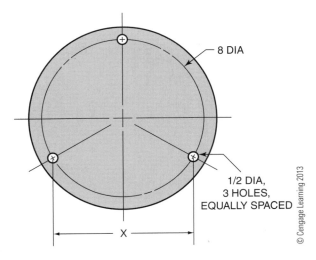

8 DIA

1/2 DIA,
3 HOLES,
EQUALLY SPACED

X

© Cengage Learning 2013

22. Calculate the length of dimension X on this template to the nearest hundredth. _____

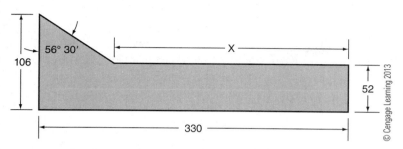

23. Determine the distance X on this match plate to the nearest thousandth. _____

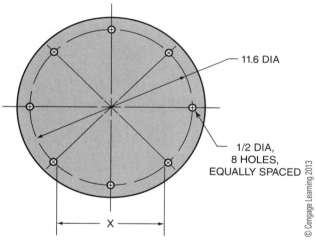

24. Calculate the distance X on this circular plate to the nearest thousandth. _____

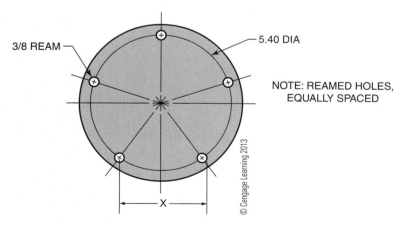

25. The drawing below is an American National thread form. Calculate
depth **D** of the thread to the nearest thousandth.   _____

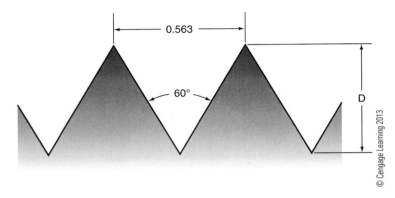

© Cengage Learning 2013

26. Calculate distance **X** on the gauge shown below to the nearest
thousandth.   _____

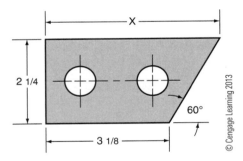

© Cengage Learning 2013

27. A drafter is designing a template. It must be rotated on pivot hole **A**
to punch the other three holes. Determine dimension **X** to the nearest
thousandth.   _____

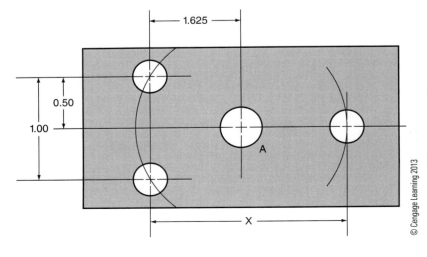

© Cengage Learning 2013

28. Calculate the distance across the crests of the acme thread shown below to the nearest thousandth.

_____

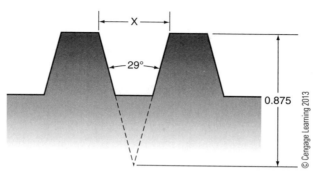

29. A drafter is designing a gauge to measure the distance between pins. Calculate dimensions **A** and **B** to the nearest thousandth

A _____

B _____

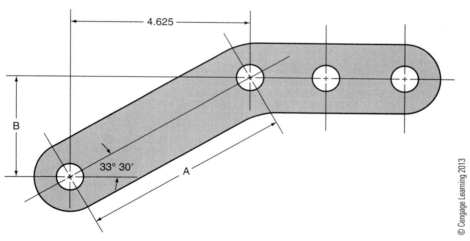

30. Determine the length **L** of the taper plug below to the nearest thousandth.

_____

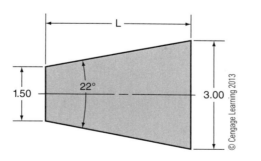

31. Calculate the inclined distance between holes **A** and **B** using the CAD drawing below. Express the answer to the nearest hundredth. _____

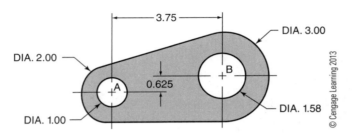

32. Determine the vertical height of the auxiliary plane **A** on the CAD drawing below. Express the answer to the nearest thousandth. _____

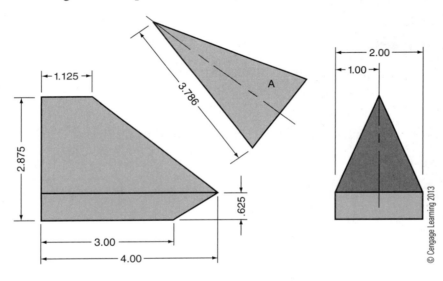

33. The measure of ∠**B** in the CAD drawing below is 30° 30′. Calculate the lengths of line **A** and line **C.** Express the answers to the nearest hundredth.

Line **A** _____

Line **C** _____

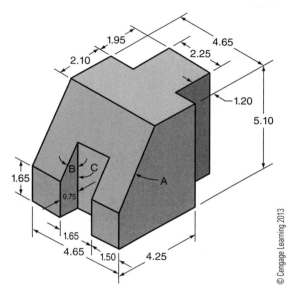

© Cengage Learning 2013

34. Use the CAD drawing below and determine distances **A** and **B** to the nearest hundredth inch.

A _____

B _____

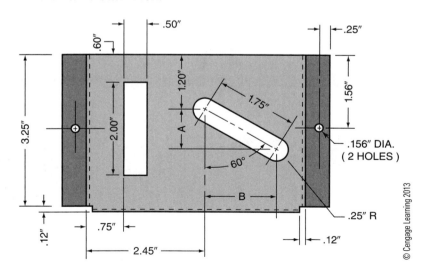

© Cengage Learning 2013

35   a. Use the CAD drawing of the locating block below and calculate the
     length of line **A** to the nearest thousandth.   a. _____

     b. Calculate the measure of ∠**B** to the nearest minute.   b. _____

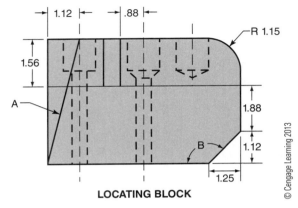

LOCATING BLOCK

36. Use the CAD drawing below and determine the overall height and
     width to the nearest hundredth.   Height _____

                                                                              Width _____

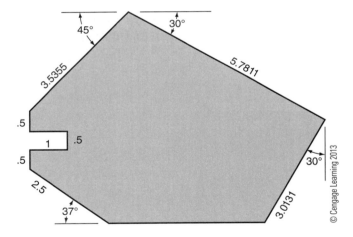

37. Use the CAD drawing below and determine the lengths of lines *A* and
B to the nearest hundredth inch.

Line *A* _____

Line *B* _____

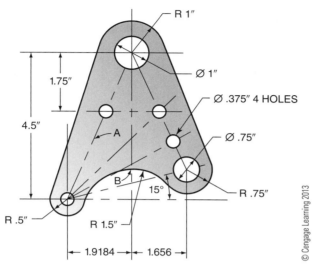

38. A drafting student is using an 8-inch 45°-45°-90° triangle to construct
a drawing. She would like to know the approximate length of the
hypotenuse. Determine the length of the hypotenuse in inches,
expressing the answer to two decimal places.

_____

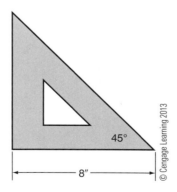

© Cengage Learning 2013

39. A drafting student measures one leg of a 30°-60°-90° triangle and finds it to be 10 inches. Determine the length, in inches, of the shorter leg and the hypotenuse. Express the answers to two decimal places.

Shorter leg _____

Hypotenuse _____

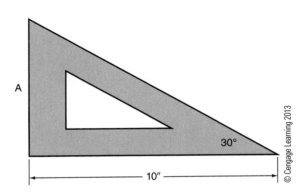

40. Determine the grid spacing for the CAD drawing below rounded to the nearest whole number.

_____

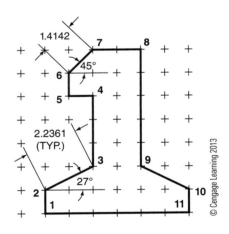

UNIT 37

Basic Principles

Oblique triangles are as important and as common to all drafters and CAD operators as right triangles. An oblique triangle is one that does not contain a right angle. The triangle can be either acute (all angles have measures less than 90°), or obtuse (one angle has a measure greater than 90°). To solve problems involving oblique triangles you need to calculate measures of sides or angles in a triangle that is not a right triangle.

The two trigonometric formulas that you can use to solve for measures in oblique triangles are called the **Law of Sines** and the **Law of Cosines**. Drafters should utilize these formulas when applicable. Each law, with examples, is described below.

The Law of Sines

The Law of Sines states that in an oblique triangle the ratios between the lengths of the sides and the sines of the angles opposite those sides are proportional. You can use the Law of Sines to solve the following types of triangle problems:

- if you know the measures of two angles and the length of the side opposite one of the angles (SAA), and you need to find the length of a side. Given AA, you can find the measure of the third angle by subtracting the sum of the measures of the two given angles from 180°.
- if you know the measures of two angles and the length of the side between the two angles (ASA), and you need to find the length of a side. (You can find the measure of the third angle by subtracting the sum of the measures of the two given angles from 180°.)

NOTE: You can also use the Law of Sines if you know the length of two sides and the measure of one angle (SSA). But sometimes there are two triangles that have these measures—an

361

acute and an obtuse triangle. So, first you need to determine how many triangles there are before applying the Law of Sines.

Given any $\triangle ABC$, the Law of Sines can be expressed in different ways: as ratios between one side and the sine of its opposite angle (on the left below); or as equal ratios between corresponding sides and the sines of their opposite angles (on the right).

$$\frac{a}{\sin A} = \frac{b}{\sin B} = \frac{c}{\sin C}$$

or   and

$$\frac{\sin A}{a} = \frac{\sin B}{b} = \frac{\sin C}{c}$$

$$\frac{a}{b} = \frac{\sin A}{\sin B}, \quad \frac{b}{c} = \frac{\sin B}{\sin C} \quad \text{or} \quad \frac{a}{c} = \frac{\sin A}{\sin C}$$

or

$$\frac{b}{a} = \frac{\sin B}{\sin A}, \quad \frac{c}{b} = \frac{\sin C}{\sin B} \quad \text{or} \quad \frac{c}{a} = \frac{\sin C}{\sin A}$$

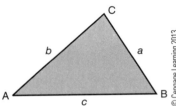

© Cengage Learning 2013

EXAMPLE: The measures of one side and two angles in $\triangle DEF$ are given below. What is X in inches to the nearest hundredth?

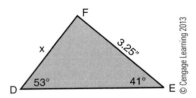

© Cengage Learning 2013

STEP 1:   The given information is two angles and a side (SAA). So, you can use the Law of Sines to find X. Start by writing a proportion that compares corresponding quantities. (Side e is represented by X.)

$$\frac{e}{\sin E} = \frac{d}{\sin D}$$

$$\frac{X}{\sin 41°} = \frac{3.25}{\sin 53°}$$

STEP 2:   Use a table of trigonometric values or a calculator to find sin 41° and sin 53° to at least four decimal places. Then substitute those values into the equation.

$$\frac{X}{0.6561} \approx \frac{3.25}{0.7986}$$

STEP 3:   Find the cross products.

$$0.7986\ X \approx 0.6571 \times 3.25$$

STEP 4:   Divide both sides of the equation by 0.7986 and solve for X.

$$X \approx \frac{0.6561 \times 3.25}{0.7986} \approx 2.67$$

So, to the nearest hundredth, e is about 2.67 inches.

3.25 ⊠ 41 ⎡SIN⎤ ⎡÷⎤ 53 ⎡SIN⎤ ⎡=⎤

The Law of Cosines

The Law of Cosines states that in a triangle, the square of the length of a side is equal to the sum of the squares of the lengths of the other sides, minus twice the product of the lengths of the two sides and the cosine of their included angle. You can use the Law of Cosines to solve the following types of problems:

- if you know the lengths of two sides and the measure of their included angle (SAS), and you need to find the length of the third side.
- if you know the lengths of the three sides (SSS), and you need to find the measure of an angle.

Given two sides and the included angle (SAS), you can use the Law of Cosines written in one of the following ways, depending on which side you need to find. When solving for the length of a side, notice that it is the cosine of the angle *opposite* that side that is part of the term on the right.

NOTE: The Law of Cosines is a modification of the Pythagorean theorem, $c^2 = a^2 + b^2$, with an extra term subtracted from the sum of the squares of the legs, $a^2 + b^2$. This may help you remember the formulas.

$$a^2 = b^2 + c^2 - 2bc \cos A$$
$$b^2 = a^2 + c^2 - 2ac \cos B$$
$$c^2 = a^2 + b^2 - 2ab \cos C$$

To use the Law of Cosines when you know the lengths of the three sides of a triangle (SSS), you can use the Law of Cosines written in one of the following ways depending on the angle you need to find.

NOTE: The term in the numerator being subtracted is always the length of the side *opposite* the angle you are solving for.

$$\cos A = \frac{b^2 + c^2 - a^2}{2bc} \qquad \cos B = \frac{a^2 + c^2 - b^2}{2ac} \qquad \cos C = \frac{a^2 + b^2 - c^2}{2ab}$$

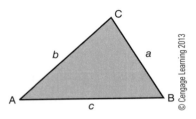

© Cengage Learning 2013

EXAMPLE: In △*ABC* below, the measures in inches of two sides and the included angle are given. What is the length in inches of side *a* to the nearest hundredth?

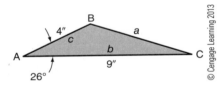

© Cengage Learning 2013

STEP 1:   The given information is two sides and the included angle (SAS). So, you can use the Law of Cosines to find *a*. Side *a* lies opposite angle *A*, and *b* = 9, *c* = 4, and ∠*A* = 26°. So the formula to use is $a^2 = b^2 + c^2 - 2\,bc\,\cos A$

$a^2 = 4^2 + 9^2 - 2(9)(4)\,\cos 26°$

STEP 2:   Use a trig table or a calculator to find cos 26°. Substitute that value into the equation.

$a^2 \approx 4^2 + 9^2 - 2(9)(4)\,(0.8988)$

Square 4 and 9.
$a^2 \approx 16 + 81 - 2(9)(4)\,(0.8988)$

Simplify the third term and round the product to three places.
$a^2 \approx 16 + 81 - 64.714$

Add 16 and 81, and then subtract 64.714.
$a^2 \approx 97 - 64.71 \approx 32.29$

STEP 3: Use a table, a calculator, or estimate the square root of 32.29 to determine a to the nearest hundredth.

$a \approx \sqrt{32.29} \approx 5.68$

So, side a is about 5.68 inches.

Finding Trig Ratios of Obtuse Angles

Two angles are *supplementary* if the sum of their measures is 180°. Each angle is said to be the *supplement* of the other. So, if one angle is obtuse, its supplement is acute, and vice versa. This is important because the cosine and tangent ratios of obtuse angles are equal to the opposite (negative) of the cosine and tangent ratios of their supplements. However, unlike the cosine and tangent ratios, the sine ratio of an obtuse angle is equal to the sine ratio of its supplement. Both are positive numbers.

To find the trig ratios of obtuse $\angle A$, use the following relationships, where $(180° - \angle A)$ is the measure of the supplement of $\angle A$.

$\cos \angle A = -\cos (180° - \angle A)$ $\tan \angle A = -\tan (180° - \angle A)$ $\sin \angle A = \sin (180° - \angle A)$

EXAMPLE: What are cos 125° and sin 125°?

The supplement of 125° is $(180° - 125°) = 55°$.

$\cos 125° = -\cos 55° \approx -0.5736$

$\sin 125° = \sin 55° \approx 0.819$.

NOTE: To find cosines and tangents of obtuse angles expressed in decimal form or in degrees, minutes, and seconds, you can use a calculator in the usual way or interpolate. (See Unit 36.) Just be sure to place a negative sign in front of the trig ratios when evaluating expressions involving cosines and tangents of obtuse angles.

Skill Problems

1. Use interpolation to find the sine of 116° 17″ and express the answer to five decimal places. (Hint: The sine of an obtuse angle equals the sine of its supplement.)

   _____

2. What is cos 147° rounded to five decimal places? (Hint: The cosine of an obtuse angle is the opposite (negative) of the cosine of its supplement.)

   _____

3. An oblique triangle contains angles of 17° 37° 23″ and 109° 26° 47″. Determine the measure of the third angle in degrees, minutes, and seconds.

   _____

4. Given a triangle whose two angles measures are 35° 34′ and 81° 27′, determine the measure of the third angle in degrees and minutes.

   _____

5. The sum of the measures of two angles of a triangle is 127° 43′ 19″. What is the measure of the third angle in degrees, minutes, and seconds?

   _____

6. What is the cofunction of the sine of 41° 19′?

   _____

7. Use interpolation to find the tangent of 142° 13′. Round the answer to five decimal places. (Hint: Interpolate between the tangents of the supplements of 142° and 143°.)

   _____

8. Use interpolation to find tan 127° 41″ rounded to four decimal places.   _____

Practical Problems

9. A mechanical drafter must determine the distance between points B and C on a detail drawing. The distance between points A and C is 1.1019 inches and between points A and B is 1.398 inches. The measure of $\angle C$ is 53°. What is the distance in inches between points B and C to the nearest thousandth? (Hint: A simple drawing is helpful.)

   _____

10. Calculate the lengths to the nearest hundredth of side a and side b in this triangle. (Hint: The given information is SAA.)

Side a _____

Side b _____

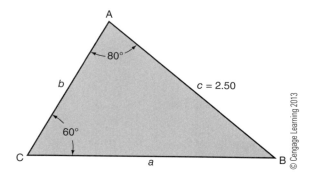

11. Given triangle **ABC** below, calculate the length of side a to the nearest thousandth. (Hint: Calculate the measure of $\angle C$ to the nearest second to find a.)

_____

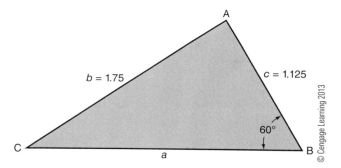

12. Calculate to the nearest hundredth side X of this triangle.

_____

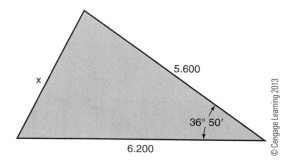

13. Calculate the measures of the three angles of this triangle to the nearest minute. (Hint: The given information is SSS.)

∠A _____

∠B _____

∠C _____

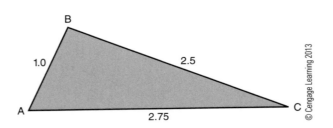

14. What is the measure to the nearest degree of obtuse ∠*X* in triangle BDC?

_____

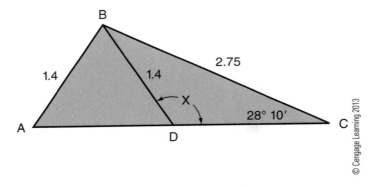

15. What is the measure to the nearest minute of ∠*B*?

_____

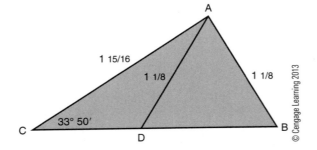

CAD Problems

16. Calculate to the nearest minute the measure of $\angle A$ in this diagram. _____

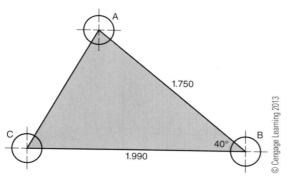

© Cengage Learning 2013

17. Calculate to the nearest hundredth distance X on this plate. _____

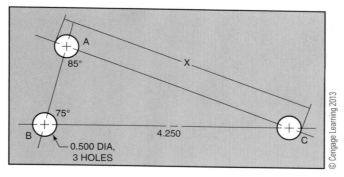

© Cengage Learning 2013

18. The distance in inches between the centers of **A** and **C** on this drawing of a circular plate is 1.625 inches. Calculate the distance in inches between the centers of holes **B** and **C**. Express the answer to four decimal places. _____

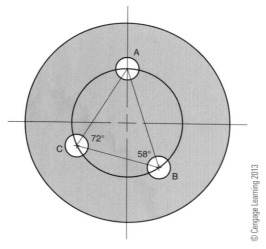

© Cengage Learning 2013

19. Calculate to the nearest thousandth dimension **X** on this template.   _____

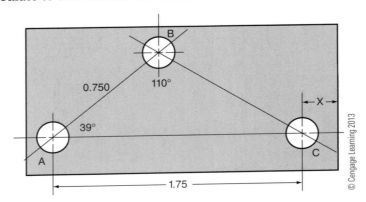

© Cengage Learning 2013

20. What is the distance between the centers of holes **C** and **A** on this layout fixture? Express the answer to the nearest hundredth.   _____

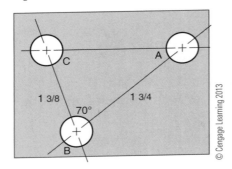

© Cengage Learning 2013

21. Determine to the nearest thousandth the lengths in inches of lines *A,* *B,* and *C* in this CAD drawing. (Hint: Use both the Law of Sines and Law of Cosines.)

Line **A** _____

Line **B** _____

Line **C** _____

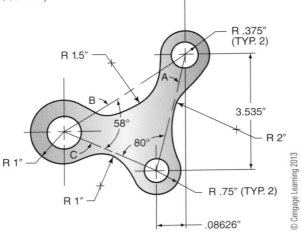

© Cengage Learning 2013

22. Use the CAD drawing below and calculate the lengths in inches of
lines **B** and **C** if line **A** is 3.536″. Express the answers to the nearest
hundredth.

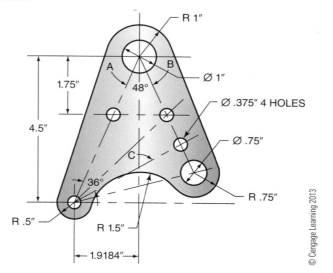

Line **B** _____

Line **C** _____

23. The measure of ∠**C** in △**ABC** the CAD drawing below is 80°. Use the
drawing and determine the length of line **AB** to the nearest hundredth. _____

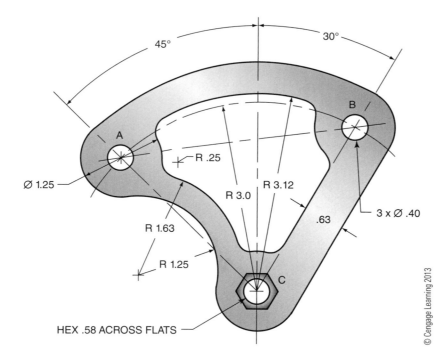

24. Use the CAD drawing below and determine the lengths in inches of lines **B** and **C** if line **A** is 3.536" long. Express the answers to the nearest thousandth.

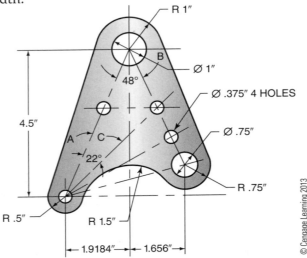

Line **B** _____

Line **C** _____

R 1"

Ø 1"

48°

Ø .375" 4 HOLES

Ø .75"

4.5"

A

C

22°

R .75"

R .5"

R 1.5"

1.9184" — 1.656"

© Cengage Learning 2013

25. Determine the length of lines **AB** and **AD** in the CAD drawing of the plate below. Express the answers to the nearest hundredth.

Line **AB** _____

Line **AD** _____

MATCH PLATE
SCALE: HALF

3.0 DIA
B.C.

5.90 DIA

4.75 DIA
B.C.

.50 DIA
TYP

22.5° 17°

45° 45°

C

D

B

A

© Cengage Learning 2013

Estimation and Tolerance

UNIT 38

Basic Principles

An *estimation,* or an *estimate,* is an approximation of an expected value. For example, estimates in drafting can be used to describe how much time may be needed to complete a task or what it might cost to buy materials.

An estimate can be greater than or less than an expected value, or even equal to it on some occasions. The *percent accuracy* of an estimate describes how close the estimate is to an expected value.

EXAMPLE: An architect estimates that it will take 56 hours to complete a proposed design for a customer. The architect completes the design in 52 hours. What was the percent accuracy of his estimate?

METHOD 1

STEP 1: To start, you can calculate the percent error of the estimate by writing the ratio of the difference between the actual and the estimated times to the estimated time. Multiply by 100 to convert the quotient to a percent.

$$\% \text{ Error } = \frac{\text{Actual} - \text{Estimate}}{\text{Estimate}} \times 100$$

$$= \frac{52 - 56}{56} \times 100$$

$$= \frac{-4}{56} \times 100$$

NOTE: The ratio is negative because the actual time was less than the estimated time.

374

STEP 2:   Express $\dfrac{-4}{56}$ as a decimal rounded to the nearest thousandth.

$$\dfrac{-4}{56} \approx -0.071$$

STEP 3:   Multiply the decimal by 100 to convert it to a percent.

$$-0.071 \times 100 = -7.1\%$$

So, the percent error was about 7% below the estimate.

STEP 4:   To find the percent accuracy of the estimate, subtract 7.1% from 100%.

$$100\%. - 7.1\% = 92.9\%$$

So, the percent accuracy was about 93% of the estimate.

▦   100 ⊟ ⦅ 56 ⊟ 52 ⦆ ⌹ 56 ⌧ 100 ⊜

NOTE: You can check the answer by calculating 93% of the estimate, 56 hr. The answer should be approximately equal to the actual time. 0.93×56 hr. ≈ 52 hr. ✓

METHOD 2

STEP 1:   Let *X* represent the unknown part of 100. Then write a proportion that shows that *X* parts of 100 are equal to 52 parts of 56, the ratio of the actual time to the estimated time.

$$\dfrac{X}{100} = \dfrac{52}{56}$$

STEP 2:   Cross-multiply. Simplify and round the answer to the nearest tenth of a percent.

$$56X = 52 \times 100$$

$$X = \frac{52}{56} \times 100$$

$$X \approx 92.9\%$$

The ratio $\dfrac{52}{56} \approx \dfrac{92.9}{100}$.

As before, the percent accuracy of the estimate is about 93%.

 52 ÷ 56 × 100 =

NOTE: You can indicate how reasonable an estimate is in either of two ways.

- You can be *within* a certain percentage of the estimate, say 7% less than the estimate.
- Or you can be a certain percentage of the estimate, say 93% of the estimate.

Practical Problems

1. An office supervisor estimates that she will spend $129.00 on supplies next month. When the bill for supplies arrives, the cost is only $112.00. To the nearest tenth, calculate the percent accuracy of the supervisor's estimate. (Hint: The actual cost was less than the estimate, so the percentage will be less than 100%.) _____

2. An architectural drafter estimates that 180 windows will be needed in a building. The drafter determines that only 165 windows are actually needed. Find the percent accuracy of the estimate to the nearest tenth. _____

3. A structural drafter estimates that 50 gusset plates will be needed to frame a building. He finds, however, that only 37 plates are needed. The actual amount is within what percent of the estimate? _____

4. A drafter estimates that a parcel of land contains 11,250 square feet. When surveyed, the parcel of land was found to contain only 10,980 square feet. Find the percent accuracy of the estimate to the nearest tenth. _____

5. A mechanical drafter estimated that it would take 128 hours to complete a project. The time for the actual project broke down as follows: electrical schematics, 16 hr.; an assembly drawing, 31 hr.; casting drawings, 26 hr.; and machined parts, 47 hr. What percent of the estimate, to the nearest whole number, was the actual amount of time spent on the project?   _____

6. A CAD drafter estimates that it will take 60 hr. to complete a drafting project. It actually took her 66 hr. to finish the project. What was the percent accuracy of the estimate? (Hint: The actual time was greater than the estimate, so the percentage will be greater than 100%.)   _____

7. An architect estimates that the square footage of a building that will be used for manufacturing would be 12,750 square feet. The owner actually uses 12,250 square feet of space for manufacturing purposes. The actual amount is within what percent of the estimate? Round the answer to the nearest whole number.   _____

8. A drafting department supervisor estimates that 500 hours of drafting will be spent during a specific month. In fact, 535 hours are logged for that month. Find the percent accuracy of the estimate.   _____

9. A CAD drafter estimates that 35 symbols were used to create a floor plan of a large building. When counted, 47 symbols were actually used. The actual amount is within what percent of the estimate, rounded to the nearest whole number?   _____

10. An architect estimates that it will take 120 bundles of roofing shingles to cover the roof of a building. When the job is completed, it is found that 116 bundles were used. What is the percent accuracy of the estimate?   _____

11. A CAD supervisor estimates that a large CAD drawing will take 70 hours to complete. The actual drawing was completed in 68 hours. To the nearest tenth, what was the percent accuracy of the estimate?   _____

12. A CAD drafter estimates that 485 CAD symbols will be used on a set of drawings. When completed, 462 symbols were actually used. To the nearest tenth, what is the percent accuracy of the estimate?   _____

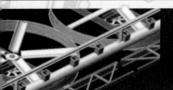

UNIT 39

Tolerancing and GD&T

Basic Principles

Tolerance

A *tolerance* is defined as a permissible variance in the size of a product. In simple language, a tolerance is the total amount that a part is permitted to vary from a specified dimension.

To calculate a tolerance, calculate the difference between the maximum (upper) dimension and the minimum (lower) dimension for a part or feature of an object.

$$\text{Tolerance} = \text{Maximum Limit} - \text{Minimum Limit}$$

Tolerancing of dimensions is necessary when the dimensions of manufactured parts must be held to a specific degree of accuracy. Interchangeable manufacturing allows parts to be made anywhere, but when assembled, they must function as designed. Tolerances and allowances placed on parts and products make this possible.

Plus-Minus Tolerance

A *plus-minus tolerance* consists of a given target value, followed by a $\pm$ (plus-minus) expression of a tolerance. You can also write this expression as the sum and difference of the target value and the $\pm$ values. This new expression is called the *limit dimension* for the tolerance.

EXAMPLE: Express $1.500 \pm .003$ as a limit dimension. Then find its tolerance.

STEP 1:   The expression $1.500 \pm .003$ written as a limit dimension is $\frac{1.503}{1.497}$.

NOTE: A limit dimension can also be written as $\frac{1.503}{1.497}$. Although it looks like a fraction, this notation is *not* a fraction. (See GD&T section that follows.)

STEP 2: The maximum (upper) limit dimension is the sum of the target number and the + value.

Upper limit = 1.500 + 0.003 = 1.503.

The minimum (lower) limit dimension is the difference between the target number and the − value.

Lower limit = 1.500 − 0.003 = 1.497.

STEP 3: The tolerance is the difference between the upper and lower limit dimensions.

Tolerance = 1.503 − 1.497 = 0.006

EXAMPLE: Express $1\frac{1}{4} \pm \frac{1}{16}$ as a limit dimension. Then find its tolerance.

STEP 1: The expression written as limit dimension is either $\frac{1\frac{5}{16}}{1\frac{3}{16}}$ or $\frac{1\frac{5}{16}}{1\frac{3}{16}}$

STEP 2: The upper limit is $1\frac{1}{4} + \frac{1}{16} = 1\frac{5}{16}$.

The lower limit is $1\frac{1}{4} - \frac{1}{16} = 1\frac{3}{16}$.

STEP 3: The tolerance is the difference between the upper and lower limits.

$$\text{Tolerance} = \left(1\frac{5}{16}\right) - \left(1\frac{3}{16}\right) = \frac{2}{16} = \frac{1}{8}$$

Bilateral Tolerance

Tolerances that vary in two directions from a given target value are called *bilateral tolerances*.

EXAMPLE: What is the tolerance for the limit dimension ± 0.006?

Subtract the lower limit, -0.006, from the upper limit, $+0.006$, to determine the tolerance.

$$\text{Tolerance} = 0.006 - (-0.006) = 0.012$$

NOTE: A positive number minus a negative number equals a positive number.

EXAMPLE: What is the tolerance for the limit dimension $^{+0.003}_{-0.002}$

Subtract the lower limit, -0.002, from the upper limit, 0.003, to determine the tolerance.

$$\text{Tolerance} = 0.003 - (-0.002) = 0.005$$

Unilateral Tolerance

Tolerances that vary in only one direction, either above or below a target value, are called *unilateral tolerances*. You can still subtract to find the tolerance.

EXAMPLE: What is the tolerance of the limit dimension $^{+0.003}_{-0.000}$?

$$\text{Tolerance} = 0.003 - (-0.00) = 0.003$$

EXAMPLE: What is the tolerance of the limit dimensions $^{+0.000}_{-0.007}$?

$$\text{Tolerance} = +0.000 - (-0.007) = 0.007$$

Allowance

In drafting, an *allowance* is the intended difference between the size of two mating parts, such as holes and shafts. The allowance is the difference between the size of the smallest hole and the size of the largest shaft. Two types of allowance fits are possible.

- A *clearance fit* provides clearance between mating parts. In this case, the allowance is always positive.
- An *interference fit* results in an interference between mating parts. In this case, the allowance is always negative.

In the two examples below, the limit dimensions for each mating part is given. Calculate the tolerances and allowances for each example. Identify the type of interference in each example.

EXAMPLE   Hole size in pulley: Limit dimension

$$\frac{\text{Smallest}}{\text{Largest}} = \frac{.500}{.502}$$

Tolerance of pulley = .502 − .500 = .002

Diameter of shaft: Limit dimension

$$\frac{\text{Largest}}{\text{Smallest}} = \frac{0.498}{0.495}$$

Tolerance of shaft = 0.498 − 0.495 = 0.003

Allowance (smallest hole size minus the largest shaft size)

Clearance fit = 0.500 − 0.498 = +0.002

NOTE: The allowance in a clearance fit is positive.

EXAMPLE   Hole size in wheel: Limit dimension

$$\frac{\text{Smallest}}{\text{Largest}} = \frac{0.750}{0.753}$$

Tolerance of wheel = 0.753 − 0.750 = 0.003

Diameter of shaft: Limit dimension

$$\frac{\text{Largest}}{\text{Smallest}} = \frac{0.758}{0.756}$$

Tolerance of shaft = 0.758 − 0.756 = 0.002

Allowance (smallest wheel size minus the largest shaft size)

Interference fit = 0.750 − 0.758 = −0.008

NOTE: The allowance in an interference fit is negative.

Geometric Dimensioning and Tolerancing (GD&T)

Over time, mechanical engineers found that the traditional ± tolerance symbols and lengthy verbal notes on drawings did not always adequately describe the amount of variation in a part. So a new symbolic drafting language, called *Geometric Dimensioning and Tolerancing (GD&T)*, was introduced. This language is now commonly used in many mechanical engineering drawings. It provides drafters with a way to clearly, precisely, and briefly describe the shape, size, and location of objects, as well as features related to their function, proximity, and relationships to each other.

GD&T provides a common language of symbols with which to communicate during the design, manufacture, assembly, and inspection of parts. This is more important than ever before, with the manufacture and assembly of parts spread across the globe.

GD&T was developed over many years and has been defined by the American National Standards Institute and the American Society of Mechanical Engineers (ANSI/ASME) in 1994. It is now an international standard.

The basic purpose of GD&T is to replace wordy and possibly ambiguous notes with a strict language of symbols and rules that offer the following advantages:

- Uniformity: Symbols that can reduce or eliminate ambiguity.
- Efficiency: Symbols that are compact and that can be positioned closer to the objects they describe.
- Universality: Symbols that transcend language barriers.
- Legibility: Symbols that are easier to read and interpret.
- Precision: Symbols that eliminate a level of error introduced.
- Precedence: Symbols that are used in many other disciplines, such as electrical and electronics schematics, welding instructions, and architectural and structural diagrams.

The two CAD diagrams that follow illustrate how the traditional system and the GD&T system can be used to specify the same object and its parts. Compare the two figures to see how the use of GD&T simplifies the specifications of all the parts of an object.

Traditional tolerance system

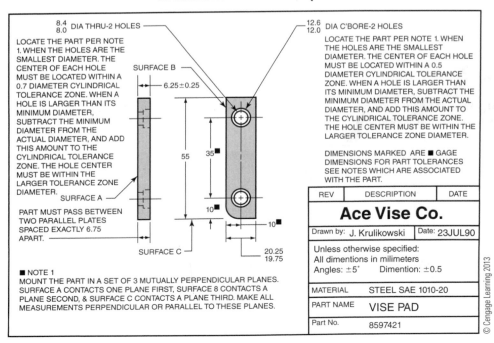

GD&T tolerance system

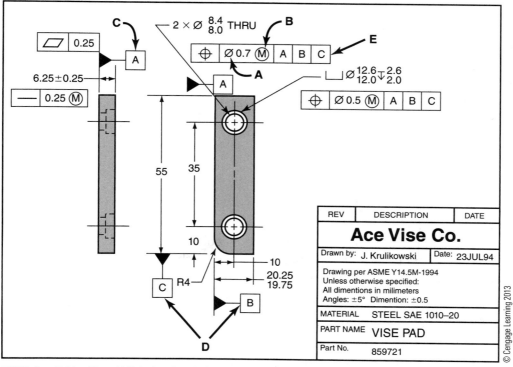

NOTE: See Tables VI and VII in Section II of the Appendix on pages 397 and 398 for descriptions of some of the GD&T symbols in the CAD figure above.

Practical Problems

1. The upper and lower dimension limits for an object are given as $^{0.861}_{0.827}$. What is the tolerance for this dimension?   _____

2. Calculate the tolerance on the dimension expressed as 1.125 ± 0.009.   _____

3. Determine the upper and lower dimension limits and tolerance of the dimensional value $4.375 \pm ^{.015}_{.009}$.

    Upper limit _____

    Lower limit _____

    Tolerance _____

4. What is the tolerance for the dimension $3\dfrac{1}{2} \begin{array}{c} +\ \dfrac{1}{16.} \\ -\ 0 \end{array}$ _____

5. Calculate the lower limit for a dimension having a bilateral tolerance of ± 0.019 and an upper limit of 2.625. _____

6. Determine the upper and lower limits for a dimension expressed as 3⁵⁄₁₆ ± ¹⁄₆₄.

   Upper limit _____

   Lower limit _____

7. Determine the upper limit for a dimension having a unilateral tolerance of $^{+0.017}_{-0.000}$ and a lower limit of 4.375. _____

8. Determine the upper and lower limits for a dimension expressed as 1½ + ¹⁄₃₂ − ¹⁄₆₄. Express the answers in simplest form.

   Upper limit _____

   Lower limit _____

9. Calculate the tolerances for a hole dimension of $^{0.6250}_{0.6262}$ and a shaft dimension of $^{0.6240}_{0.6228}$.

   Hole tolerance _____

   Shaft tolerance _____

10. Calculate the hole tolerance, the shaft tolerance, and the allowance for a $^{2.2496}_{2.2464}$ hole diameter and a shaft diameter of $^{2.2485}_{2.2464}$?

    Hole tolerance _____

    Shaft tolerance _____

    Allowance _____

11. A limit dimension on a drawing is given as $^{3.15062}_{3.01057}$. What is the deviation in this dimension? _____

12. Determine the allowance for a hole dimension of 1.625 $^{+0.002}_{-0.005}$ and a shaft dimension of 1.635 $^{+0.009}_{-0.002}$. _____

13. The dimension of a hole is 1.500 ± 0.003, and the allowance is 0.007. The tolerance on the shaft is 0.005. Calculate the limits for the shaft if the result is a clearance fit. _____

14. A shaft dimension is 1.375 $^{+0.009}_{-0.003}$, and the allowance is 0.027. The tolerance on the hole is 0.009. Calculate the limits for the hole if the result is an interference fit. _____

CAD Problems

15. Calculate the tolerances for the holes contained on the CAD drawing of the shim below. Express **A** as a decimal and **B** as a fraction in simplest form.

A _____

B _____

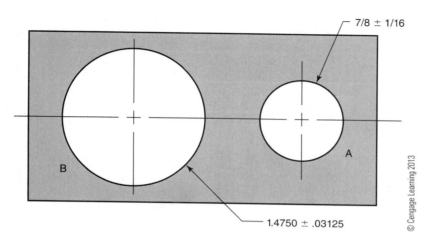

7/8 ± 1/16

A

B

1.4750 ± .03125

© Cengage Learning 2013

16. Calculate the hole and shaft tolerances and the allowance for the parts in the CAD drawings below. Then identify the type of fit this is.

Hole tolerance _____

Shaft tolerance _____

Allowance _____

Type of fit _____

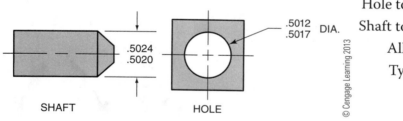

.5024
.5020

SHAFT

.5012
.5017   DIA.

HOLE

© Cengage Learning 2013

17. Determine the upper and lower limits and the allowance for the CAD drawings below. What type of fit results when these two parts are assembled?

Upper limit _____

Lower limit _____

Allowance _____

Type of fit _____

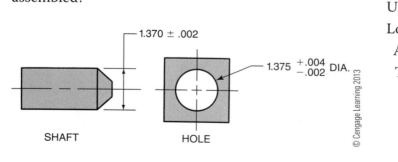

1.370 ± .002

SHAFT

$1.375 \begin{smallmatrix} +.004 \\ -.002 \end{smallmatrix}$ DIA.

HOLE

© Cengage Learning 2013

18. Use the CAD drawings below and determine the tolerances applied to dimensions **A** and **B**. Assume that dimension **A** is at its largest size. What will be the internal diameter for the part that will mate with **A** and also provide an allowance of 0.0037"?

Tolerance **A** _____

Tolerance **B** _____

Internal diameter _____

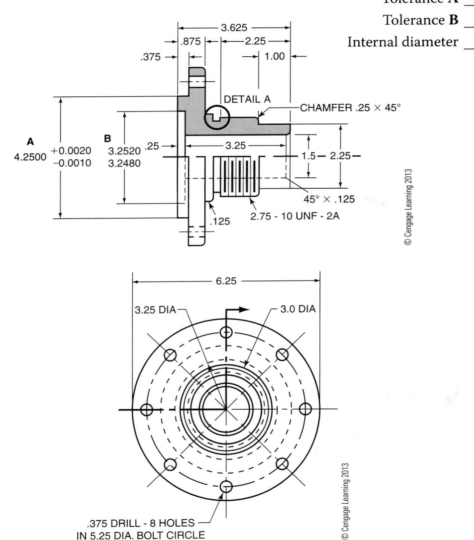

19. Determine the hole tolerance, the shaft tolerance, and the allowance
for the mating parts in the CAD drawings below.

Hole tolerance _____

Shaft tolerance _____

Allowance _____

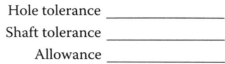

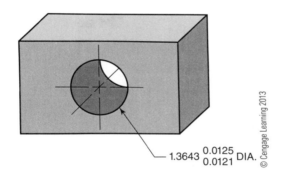

1.7500
1.7491

1.7478
1.7488

© Cengage Learning 2013

20. a. What is the *smallest* allowable size for the hole in this CAD drawing?   a. _____

b. What is the *largest* allowable size for the hole in this drawing?   b. _____

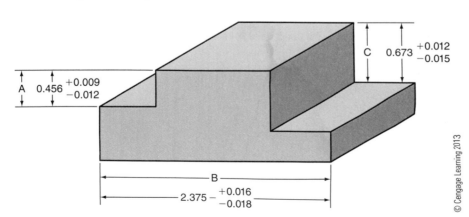

1.3643 $^{0.0125}_{0.0121}$ DIA.

© Cengage Learning 2013

21. What is the *smallest* permissible size for each dimension on this CAD
drawing of a set-up block?

A _____

B _____

C _____

C   0.673 $^{+0.012}_{-0.015}$

A   0.456 $^{+0.009}_{-0.012}$

B

2.375 $^{+0.016}_{-0.018}$

© Cengage Learning 2013

22. The CAD drawing below shows the basic sizes and tolerances of a shim.
Calculate the *largest* length (A) and width (B) permitted.

A _____

B _____

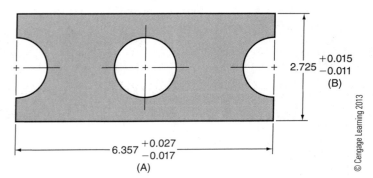

$2.725 \begin{smallmatrix} +0.015 \\ -0.011 \end{smallmatrix}$
(B)

$6.357 \begin{smallmatrix} +0.027 \\ -0.017 \end{smallmatrix}$
(A)

APPENDIX

Section I The Scientific Calculator

There are numerous brands and types of scientific calculators that you can purchase to help you deal with many mathematical calculations. These calculations involve squaring numbers, finding powers and roots, evaluating trigonometric ratios and so on. A list of the common keys found on most scientific calculators and used most often in drafting appears below.

KEY	FUNCTION
ON/C	Turns the calculator on or clears the display
OFF	Turns the calculator off
2nd	Accesses the function that appears above a key
SCI	Sets display to scientific notation
ENG	Sets display to engineering notation
DRG	Sets angle measures in degrees (DEG), radians (RAD), or gradient measure (GRAD)
(	Inserts a left parenthesis
)	Inserts a right parenthesis
FIX	Fixes the number of significant digits
STO	Stores the displayed value in memory

RCL	Recalls the value stored in memory
ab/c	Enters fractions or mixed numbers
F ⟷ D	Toggles the number in the display between fractional and decimal form
d/c	Toggles the number in the display between a mixed number and an improper fraction
+	Performs addition
−	Performs subtraction
×	Performs multiplication
÷	Performs division
=	Completes all pending operations
+/−	Changes the sign of the displayed number
x!	Displays the factorial of a number
π	Displays an approximation of pi to several decimal places
→	Removes the last digit from a number entry
%	Displays a number as the decimal equivalent of a percent
x^2	Squares a number
x^3	Cubes a number
y^x	Displays a number y raised to the power of x
$\sqrt[x]{}$	Displays the root of a number indicated by the index, x

Key	Description
$\sqrt{x}$	Displays the square root of a non-negative number
$\sqrt[3]{}$	Displays the cube root of a number
1/x	Displays the reciprocal of the displayed value (x not equal to 0)
LOG	Displays the common logarithm of a positive number
DMS ► DD	Converts an angle measure in degrees, minutes, and seconds to a decimal
DD ► DMS	Converts a degree measure in decimal form into degrees, minutes, and seconds
SIN	Displays the sine of an angle
COS	Displays the cosine of an angle
TAN	Displays the tangent of an angle
CSR	Clears any data values stored in memory (register)
Σ^+	Adds a value to a set of data values stored in memory
Σ^-	Subtracts a value from a set of data values stored in memory
Σx	Adds the set of data values stored in memory
n	Displays how many data values are stored in memory
$\bar{x}$	Displays the mean (average) of a set of data values stored in memory

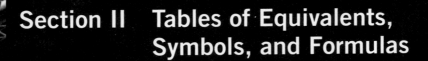

Section II Tables of Equivalents, Symbols, and Formulas

TABLE 1
EQUIVALENT U.S. CUSTOMARY UNITS AND METRIC UNITS OF MEASURE

Linear Measure

Unit	Inches to millimetres	Millimetres to inches	Feet to metres	Metres to feet	Yards to metres	Metres to yards	Miles to kilometres	Kilometres to miles
1	25.40	0.03937	0.3048	3.281	0.9144	1.094	1.609	0.6214
2	50.80	0.07874	0.6096	6.562	1.829	2.187	3.219	1.243
3	76.20	0.1181	0.9144	9.842	2.743	3.281	4.828	1.864
4	101.60	0.1575	1.219	13.12	3.658	4.374	6.437	2.485
5	127.00	0.1968	1.524	16.40	4.572	5.468	8.047	3.107
6	152.40	0.2362	1.829	19.68	5.486	6.562	9.656	3.728
7	177.80	0.2756	2.134	22.97	6.401	7.655	11.27	4.350
8	203.20	0.3150	2.438	26.25	7.315	8.749	12.87	4.971
9	228.60	0.3543	2.743	29.53	8.230	9.842	14.48	5.592

Example 1 in. = 25.40 mm, 1 m = 3.281 ft., 1 km = 0.6214 mi.

Surface Measure

Unit	Square inches to square centimetres	Square centimetres to square inches	Square feet to square metres	Square metres to square feet	Square yards to square metres	Square metres to square yards	Acres to hectares	Hectares to acres	Square miles to square kilometres	Square kilometres to square miles
1	6.452	0.1550	0.0929	10.76	0.8361	1.196	0.4047	2.471	2.59	0.3861
2	12.90	0.31	0.1859	21.53	1.672	2.392	0.8094	4.942	5.18	0.7722
3	19.356	0.465	0.2787	32.29	2.508	3.588	1.214	7.413	7.77	1.158
4	25.81	0.62	0.3716	43.06	3.345	4.784	1.619	9.884	10.36	1.544
5	32.26	0.775	0.4645	53.82	4.181	5.98	2.023	12.355	12.95	1.931
6	38.71	0.93	0.5574	64.58	5.017	7.176	2.428	14.826	15.54	2.317
7	45.16	1.085	0.6503	75.35	5.853	8.372	2.833	17.297	18.13	2.703
8	51.61	1.24	0.7432	86.11	6.689	9.568	3.237	19.768	20.72	3.089
9	58.08	1.395	0.8361	96.87	7.525	10.764	3.642	22.239	23.31	3.475

Example 1 sq. in. = 6.452 cm^2, 1 m^2 = 1.196 sq. yd., 1 sq. mi. = 2.59 km^2

Cubic Measure

Unit	Cubic inches to cubic centimetres	Cubic centimetres to cubic inches	Cubic feet to cubic metres	Cubic metres to cubic feet	Cubic yards to cubic metres	Cubic metres to cubic yards	Gallons to cubic feet	Cubic feet to gallons
1	16.39	0.06102	0.02832	35.31	0.7646	1.308	0.1337	7.481
2	32.77	0.1220	0.05663	70.63	1.529	2.616	0.2674	14.96
3	49.16	0.1831	0.08495	105.9	2.294	3.924	0.4010	22.44
4	65.55	0.2441	0.1133	141.3	3.058	5.232	0.5347	29.92
5	81.94	0.3051	0.1416	176.6	3.823	6.540	0.6684	37.40
6	98.32	0.3661	0.1699	211.9	4.587	7.848	0.8021	44.88
7	114.7	0.4272	0.1982	247.2	5.352	9.156	0.9358	52.36
8	131.1	0.4882	0.2265	282.5	6.116	10.46	1.069	59.84
9	147.5	0.5492	0.2549	371.8	6.881	11.77	1.203	67.32

Example 1 cm^3 = 0.06102 cu. in., 1 gal. = 0.1337 cu. ft.

Volume or Capacity Measure

Unit	Liquid ounces to cubic centimetres	Cubic centimetres to liquid ounces	Pints to litres	Litres to pints	Quarts to litres	Litres to quarts	Gallons to litres	Litres to gallons	Bushels to hectolitres	Hectolitres to bushels
1	29.57	0.03381	0.4732	2.113	0.9463	1.057	3.785	0.2642	0.3524	2.838
2	59.15	0.06763	0.9463	4.227	1.893	2.113	7.571	0.5284	0.7048	5.676
3	88.72	0.1014	1.420	6.340	2.839	3.785	11.36	0.7925	1.057	8.513
4	118.3	0.1353	1.893	8.454	3.170	4.227	15.14	1.057	1.410	11.35
5	147.9	0.1691	2.366	10.57	4.732	5.284	18.93	1.321	1.762	14.19
6	177.4	0.2029	2.839	12.68	5.678	6.340	22.71	1.585	2.114	17.03
7	207.0	0.2367	3.312	14.79	6.624	7.397	26.50	1.849	2.467	19.86
8	236.6	0.2705	3.785	16.91	7.571	8.454	30.28	2.113	2.819	22.70
9	266.2	0.3043	4.259	19.02	8.517	9.510	34.07	2.378	3.171	25.54

Example 1L = 2.113 pt., 1 gal. = 3.785 L

TABLE II
FRACTION AND DECIMAL EQUIVALENTS

Fraction	Decimal Equivalent		Fraction	Decimal Equivalent	
	Customary (in.)	Metric (mm)		Customary (in.)	Metric (mm)
1/64 — .015625		0.3969	33/64 — .515625		13.0969
1/32 — .03125		0.7938	17/32 — .53125		13.4938
3/64 — .046875		1.1906	35/64 — .546875		13.8906
1/16 — .0625		1.5875	9/16 — .5625		14.2875
5/64 — .078125		1.9844	37/64 — .578125		14.6844
3/32 — .09375		2.3813	19/32 — .59375		15.0813
7/64 — .109375		2.7781	39/64 — .609375		15.4781
1/8 — .1250		3.1750	5/8 — .6250		15.8750
9/64 — .140625		3.5719	41/64 — .640625		16.2719
5/32 — .15625		3.9688	21/32 — .65625		16.6688
11/64 — .171875		4.3656	43/64 — .671875		17.0656
3/16 — .1875		4.7625	11/16 — .6875		17.4625
13/64 — .203125		5.1594	45/64 — .703125		17.8594
7/32 — .21875		5.5563	23/32 — .71875		18.2563
15/64 — .234375		5.9531	47/64 — .734375		18.6531
1/4 — .250		6.3500	3/4 — .750		19.0500
17/64 — .265625		6.7469	49/64 — .765625		19.4469
9/32 — .28125		7.1438	25/32 — .78125		19.8438
19/64 — .296875		7.5406	51/64 — .796875		20.2406
5/16 — .3125		7.9375	13/16 — .8125		20.6375
21/64 — .328125		8.3384	53/64 — .828125		21.0344
11/32 — .34375		8.7313	27/32 — .84375		21.4313
23/64 — .359375		9.1281	55/64 — .859375		21.8281
3/8 — .3750		9.5250	7/8 — .8750		22.2250
25/64 — .390625		9.9219	57/64 — .890625		22.6219
13/32 — .40625		10.3188	29/32 — .90625		23.0188
27/64 — .421875		10.7156	59/64 — .921875		23.4156
7/16 — .4375		11.1125	15/16 — .9375		23.8125
29/64 — .453125		11.5094	61/64 — .953125		24.2094
15/32 — .46875		11.9063	31/32 — .96875		24.6063
31/64 — .484375		12.3031	63/64 — .984375		25.0031
1/2 — .500		12.7000	1 — 1.000		25.4000

TABLE III
POWERS AND ROOTS OF NUMBERS (1 through 100)

Number	Powers		Roots		Number	Powers		Roots	
	Square	Cube	Square	Cube		Square	Cube	Square	Cube
1	1	1	1.000	1.000	51	2,601	132,651	7.141	3.708
2	4	8	1.414	1.260	52	2,704	140,608	7.211	3.733
3	9	27	1.732	1.442	53	2,809	148,877	7.280	3.756
4	16	64	2.000	1.587	54	2,916	157,464	7.348	3.780
5	25	125	2.236	1.710	55	3,025	166,375	7.416	3.803
6	36	216	2.449	1.817	56	3,136	175,616	7.483	3.826
7	49	343	2.646	1.913	57	3,249	185,193	7.550	3.849
8	64	512	2.828	2.000	58	3,364	195,112	7.616	3.871
9	81	729	3.000	2.080	59	3,481	205,379	7.681	3.893
10	100	1,000	3.162	2.154	60	3,600	216,000	7.746	3.915
11	121	1,331	3.317	2.224	61	3,721	226,981	7.810	3.936
12	144	1,728	3.464	2.289	62	3,844	238,328	7.874	3.958
13	169	2,197	3.606	2.351	63	3,969	250,047	7.937	3.979
14	196	2,744	3.742	2.410	64	4,096	262,144	8.000	4.000
15	225	3,375	3.873	2.466	65	4,225	274,625	8.062	4.021
16	256	4,096	4.000	2.520	66	4,356	287,496	8.124	4.041
17	289	4,913	4.123	2.571	67	4,489	300,763	8.185	4.062
18	324	5,832	4.243	2.621	68	4,624	314,432	8.246	4.082
19	361	6,859	4.359	2.668	69	4,761	328,509	8.307	4.102
20	400	8,000	4.472	2.714	70	4,900	343,000	8.367	4.121
21	441	9,261	4.583	2.759	71	5,041	357,911	8.426	4.141
22	484	10,648	4.690	2.802	72	5,184	373,248	8.485	4.160
23	529	12,167	4.796	2.844	73	5,329	389,017	8.544	4.179
24	576	13,824	4.899	2.884	74	5,476	405,224	8.602	4.198
25	625	15,625	5.000	2.924	75	5,625	421,875	8.660	4.217
26	676	17,576	5.099	2.962	76	5,776	438,976	8.718	4.236
27	729	19,683	5.196	3.000	77	5,929	456,533	8.775	4.254
28	784	21,952	5.292	3.037	78	6,084	474,552	8.832	4.273
29	841	24,389	5.385	3.072	79	6,241	493,039	8.888	4.291
30	900	27,000	5.477	3.107	80	6,400	512,000	8.944	4.309
31	961	29,791	5.568	3.141	81	6,561	531,441	9.000	4.327
32	1,024	32,798	5.657	3.175	82	6,724	551,368	9.055	4.344
33	1,089	35,937	5.745	3.208	83	6,889	571,787	9.110	4.362
34	1,156	39,304	5.831	3.240	84	7,056	592,704	9.165	4.380
35	1,225	42,875	5.916	3.271	85	7,225	614,125	9.220	4.397
36	1,296	46,656	6.000	3.302	86	7,396	636,056	9.274	4.414
37	1,369	50,653	6.083	3.332	87	7,569	658,503	9.327	4.481
38	1,444	54,872	6.164	3.362	88	7,744	681,472	9.381	4.448
39	1,521	59,319	6.245	3.391	89	7,921	704,969	9.434	4.465
40	1,600	64,000	6.325	3.420	90	8,100	729,000	9.487	4.481
41	1,681	68,921	6.403	3.448	91	8,281	753,571	9.539	4.498
42	1,764	74,088	6.481	3.476	92	8,464	778,688	9.592	4.514
43	1,849	79,507	6.557	3.503	93	8,649	804,357	9.644	4.531
44	1,936	85,184	6.633	3.530	94	8,836	830,584	9.695	4.547
45	2,025	91,125	6.708	3.557	95	9,025	857,375	9.747	4.563
46	2,116	97,336	6.782	3.583	96	9,216	884,736	9.798	4.579
47	2,209	103,823	6.856	3.609	97	9,409	912,673	9.849	4.595
48	2,304	110,592	6.928	3.634	98	9,604	941,192	9.900	4.610
49	2,401	117,649	7.000	3.659	99	9,801	970,299	9.950	4.626
50	2,500	125,000	7.071	3.684	100	10,000	1,000,000	10.000	4.642

TABLE IV
CIRCUMFERENCES AND AREAS (0.2 to 9.8; 10 to 99)

Diameter	Circum.	Area	Diameter	Circum.	Area
0.2	0.628	0.0314	31	97.39	754.8
0.4	1.26	0.1256	32	100.5	804.2
0.6	1.88	0.2827	33	103.7	855.3
0.8	2.51	0.5026	34	106.8	907.9
1	3.14	0.7854	35	110	962.1
1.2	3.77	1.131	36	113.1	1,017.9
1.4	4.39	1.539	37	116.2	1,075.2
1.6	5.02	2.011	38	119.4	1,134.1
1.8	5.65	2.545	39	122.5	1,194.6
2	6.28	3.142	40	125.7	1,256.6
2.2	6.91	3.801	41	128.8	1,320.3
2.4	7.53	4.524	42	131.9	1,385.4
2.6	8.16	5.309	43	135.1	1,452.2
2.8	8.79	6.158	44	138.2	1,520.5
3	9.42	7.069	45	141.4	1,590.4
3.2	10.05	7.548	46	144.5	1,661.9
3.4	10.68	8.553	47	147.7	1,734.9
3.6	11.3	10.18	48	150.8	1,809.6
3.8	11.93	11.34	49	153.9	1,885.7
4	12.57	12.57	50	157.1	1,963.5
4.2	13.19	13.85	51	160.2	2,042.8
4.4	13.82	15.21	52	163.4	2,123.7
4.6	14.45	16.62	53	166.5	2,206.2
4.8	15.08	18.1	54	169.6	2,290.2
5	15.7	19.63	55	172.8	2,375.8
5.2	16.33	21.24	56	175.9	2,463
5.4	16.96	22.9	57	179.1	2,551.8
5.6	17.59	24.63	58	182.2	2,642.1
5.8	18.22	26.42	59	185.4	2,734
6	18.84	28.27	60	188.5	2,827.4
6.2	19.47	30.19	61	191.6	2,922.5
6.4	20.1	32.17	62	194.8	3,019.1
6.6	20.73	34.21	63	197.9	3,117.3
6.8	21.36	36.32	64	201.1	3,217
7	21.99	38.48	65	204.2	3,318.3
7.2	22.61	40.72	66	207.3	3,421.2
7.4	23.24	43.01	67	210.5	3,525.7
7.6	23.87	45.36	68	213.6	3,631.7
7.8	24.5	47.78	69	216.8	3,739.3
8	25.13	50.27	70	219.9	3,848.5
8.2	25.76	52.81	71	223.1	3,959.2
8.4	26.38	55.42	72	226.2	4,071.5
8.6	27.01	58.09	73	229.3	4,185.4
8.8	27.64	60.82	74	232.5	4,300.8
9	28.27	63.62	75	235.6	4,417.9
9.2	28.9	66.48	76	238.8	4,536.5
9.4	29.53	69.4	77	241.9	4,656.6
9.6	30.15	72.38	78	245	4,778.4
9.8	30.78	75.43	79	248.2	4,901.7
10	31.41	78.54	80	251.3	5,026.6
11	34.55	95.03	81	254.5	5,153
12	37.69	113	82	257.6	5,281
13	40.84	132.7	83	260.8	5,410.6
14	43.98	153.9	84	263.9	5,541.8
15	47.12	176.7	85	267.0	5,674.5
16	50.26	201	86	270.2	5,808.8
17	53.4	226.9	87	273.3	5,944.7
18	56.54	254.4	88	276.5	6,082.1
19	59.69	283.5	89	279.6	6,221.2
20	62.83	314.1	90	282.7	6,361.7
21	65.97	346.3	91	285.9	6,503.8
22	69.11	380.1	92	289.0	6,647.6
23	72.25	415.4	93	292.2	6,792.9
24	75.39	452.3	94	295.2	6,939.8
25	78.54	490.8	95	298.5	7,088.2
26	81.68	530.9	96	301.6	7,238.2
27	84.82	572.5	97	304.7	7,389.8
28	87.96	615.7	98	307.9	7,543.0
29	91.1	660.5	99	311.9	7,697.7
30	94.24	706.8			

TABLE V
TAP DRILL SIZES FOR AMERICAN NATIONAL FORM THREADS
(U.S. CUSTOMARY AND METRIC DRILL SIZES)

Diam. of Thread	Threads per Inch	Drill	Decimal Equiv.	Diam. of Thread	Threads per Inch	Drill	Decimal Equiv.
No. 0—.060	80 NF	3/64	0.0469		12 N	39/64	0.6094
1—.073	64 NC	1.5 mm	0.0591	11/16	24 NEF	16.5 mm	0.6496
	72 NF	53	0.0595		10 NC	16.5 mm	0.6496
2—.086	56 NC	50	0.0700		12 N	17 mm	0.6693
	64 NF	50	0.0700	3/4	16 NF	17.5 mm	0.6890
3—.099	48 NC	5/64	0.0781		20 NEF	45/64	0.7031
	56 NF	45	0.0820		12 N	18.5 mm	0.7283
4—.112	40 NC	43	0.0890	13/16	16 N	3/4	0.7500
	48 NF	42	0.0935		20 NEF	49/64	0.7656
5—.125	40 NC	38	0.1015		9 NC	49/64	0.7656
	44 NF	37	0.1040		12 N	20 mm	0.7874
6—.138	32 NC	36	0.1065	7/8	14NF	20.5 mm	0.8071
	40 NF	33	0.1130		16 N	13/16	0.8125
8—.164	32 NC	29	0.1360		20 NEF	21 mm	0.8268
	36 NF	29	0.1360		12 N	55/64	0.8594
10—.190	24 NC	25	0.1495	15/16	16 N	7/8	0.8750
	32 NF	21	0.1590		20 NEF	22.5 mm	0.8858
12—.216	24 NC	16	0.1770		8 NC	7/8	0.8750
	28 NF	14	0.1820		12 N	59/64	0.9219
	20 NC	7	0.2010	1	14 NF	23.5 mm	0.9252
1/4	28 NF	3	0.2130		16 N	15/16	0.9375
	32 NEF	7/32	0.2188		20 NEF	61/64	0.9531
	18 NC	F	0.2570		6 NC	1 21/64	1.3281
5/16	24 NF	I	0.2720		8 N	1 3/8	1.3750
	32 NEF	9/32	0.2812	1 1/2	12 NF	36 mm	1.4173
	16 NC	5/16	0.3125		16N	1 7/16	1.4375
3/8	24 NF	Q	0.3320		18 NEF	1 29/64	1.4531
	32 NEF	11/32	0.3438		4 1/2 NC	1 25/32	1.7812
	14 NC	U	0.3680	2	8 N	1 7/8	1.8750
7/16	20 NF	25/64	0.3906		12 N	1 59/64	1.9219
	28 NEF	Y	0.4040		16 NEF	1 15/16	1.9375
	12 N	27/64	0.4219		4 NC	2 1/4	2.2500
1/2	13 NC	27/64	0.4219	2 1/2	8 N	2 3/8	2.3750
	20 NF	29/64	0.4531		12 N	61.5 mm	2.4213
	28 NEF	15/32	0.4687		16 N	2 7/16	2.4375
	12 NC	31/64	0.4844		4 NC	2 3/4	2.7500
9/16	18 NF	33/64	0.5156	3	8 N	2 7/8	2.8750
	24 NEF	33/64	0.5156		12 N	74 mm	2.9134
	11 NC	17/32	0.5312		16 N	2 15/16	2.9375
5/8	12 N	35/64	0.5469				
	18 NF	14.5 mm	0.5709				
	24 NEF	37/64	0.5781				

GLOSSARY

Acute angle An angle whose measure is greater than 0° and less than 90°.

Allowance Minimum clearance or maximum interference between mating parts.

Alloy A mixture of two or more metals.

Angle A figure formed by the intersection of two lines at a point called the *vertex* of the angle; the lines are called the *sides* of the angle.

Arc Curved part of a circle.

Architect A person who designs and oversees the construction of buildings.

Architectural drafter A person who prepares all types of architectural drawings and documents.

Area A number that is a measure of the amount of surface in the interior of a planar figure.

Arrowheads A symbol at one end of a dimension line or a leader that points to a feature of an object or part.

Assembly drawing A drawing showing all the parts of an object in their proper locations.

Basic size The size of an object from which tolerances are applied.

Bearing A supporting member for a rotating shaft.

Bevel An inclined edge, not at a right angle to a joining surface.

Bisect To divide an object into two equal parts.

Blueprint A reproduction of a drawing made up of white lines on a bright blue background.

Bolt circle An imaginary circle on a circular part of an object that indicates the location of bolts on the part.

Bore To enlarge a hole with a boring bar or other machine tool.

Cam A rotating member used to change circular motion to reciprocating motion.

Casting An object made by pouring molten metal into a mold.

Center line A thin line representing the center of a hole or the axis of a cylindrical or conical object.

Chamfer A slight bevel on the end of a shaft or corner of an object to avoid a sharp edge.

Chord A line segment between two points on a circle.

Circle A set of points in a plane equidistant from a center point.

Circumference The distance around a circle; its perimeter.

Circumscribe To draw a polygon or circle around another so that each vertex of the interior polygon touches the outer figure at exactly one point, or the interior circle is tangent to the sides of the outer polygon.

Civil drafter A person who prepares drawings dealing with highways, bridges, tunnels, and so on.

Clearance fit Class of fit in which clearance is always maintained between mating parts.

Collar A round flange or ring fitted to a shaft to prevent sliding.

CAD The mnemonic (memory device) that stands for Computer-Aided Drafting or Computer-Aided Design, the use of computers to draw objects.

Concave A curved depression in the surface of an object.

Concentric Two or more circles that have the same center but unequal radii.

Cone A three-dimensional object with a curved side that tapers uniformly from a circular or elliptical base to a point called its *vertex*.

Conical Shaped like a cone.

Convex An exterior rounded surface of an object.

Core To form a hollow area in a part to be cast in a mold.

Cosine (cos) In a right triangle, the ratio of the length of the leg adjacent to an acute angle to the length of the hypotenuse.

Counterbore To enlarge the end of a hole cylindrically to a given depth.

Countersink To enlarge the end of a hole conically to a specified angle.

Crosshatching Lines added to a sectional drawing to indicate the material of the part.

Cube a rectangular prism having six congruent square faces.

Cylinder A three-dimensional object having a curved side and two congruent circular or elliptical bases.

Detail drawing The drawing of a single part including its exact dimensions.

Development A drawing of the surface of an object unfolded or rolled out on a plane.

Diagonal A line segment that connects two non-adjacent vertices in a polygon.

Diameter A chord of a circle that passes through the center of the circle.

Dimension line The line between extension lines that contains the measure of a dimension.

Dowel A cylindrical pin used primarily to prevent sliding between two contacting surfaces.

Drafter A person who works with original designs to make production and manufacturing drawings.

Drill To cut a cylindrical hole using a drill bit.

Eccentric A deviation from a common center.

Edge The intersection of two faces of a polyhedron.

Ellipse An oval closed-curve formed by the intersection of a cone and a plane.

Engineer A person who solves technical problems.

Entity A point, line, arc, circle, or any object used to generate CAD drawings.

Equilateral triangle A triangle having three equal sides.

Erasing shield A drafting aid used to protect specific features when erasing.

Exponent The power of a number.

Extension lines Thin lines drawn from object lines to indicate where linear dimensions begin and end.

Face A flat surface of a polyhedron.

Fillet The inside rounded corner on a casting.

Fit The degree of looseness or tightness between two mating parts.

Flange A relatively thin rim around a part.

Floor plan The top view of a building at a specified floor level.

Full-section A view of an object that is cut in half by a cutting-plane that extends entirely through an object showing the front half removed.

Galvanize To cover a surface with a molten alloy to prevent rusting.

Gasket A thin piece of material placed between two surfaces to make a tight joint.

Graphic symbols Symbolic representations used to simplify properties and relationships.

Half-section A view of one half of a full-section of an object that removes an imaginary quarter of the object.

Hexagon A six-sided polygon.

Hidden line A thin dashed line showing a surface that is hidden behind other surfaces.

Hypotenuse The side opposite the right angle in a right triangle.

Inclined An angle between two lines or planes.

Inscribe To draw a polygon or circle inside another so that each vertex of the interior polygon touches the outer figure at exactly one point, or the interior circle is tangent to the sides of the outer polygon.

Interference fit Class of fit in which interference is always maintained between mating parts.

Isometric drawing A pictorial drawing showing a three-dimensional view of an object in two dimensions. An object's vertical lines are drawn vertically, and its horizontal lines in the width and depth planes are drawn at angles of 30° to the horizontal.

Isosceles triangle A triangle having two equal sides.

Key A small piece of metal fitting in both the shaft and hub to prevent circumferential (circular) motion.

Keyway A slot in a hub or portion surrounding a shaft to receive a key.

Lathe A machine used to shape material by rotation against a tool.

Leader A line drawn from a descriptive note or symbol to an object to which the note or symbol applies.

Level Transparent layer where entities are generated and recorded on CAD drawings.

Limit dimensioning Dimensioning system in which classes of fits are controlled by applying tolerances and allowances.

Line In geometry, a one-dimensional set of points having no thickness and extending infinitely in opposite directions. In drafting, lines are shown as line segments having measurable lengths.

Mean The statistical average equal to the sum of the numeric values in a data set divided by the number of values in the set.

Mechanical drafting The drafting of drawings dealing with primarily mechanical objects.

Micrometer A precision instrument used to measure the thickness (often, the diameter) of an object.

Midpoint The point on a line segment that bisects the segment.

Nominal size The specified size of an object, which may not be its actual size.

Oblique drawing A pictorial drawing showing one face of an object parallel to the viewer.

Obtuse angle An angle whose measure is greater than 90° and less than 180°.

Obtuse triangle A triangle that has one obtuse angle.

Octagon An eight-sided polygon.

Orthographic projection A method used to represent three-dimensional objects using standard views.

Parallel Two or more lines or planes that do not intersect.

Parallelogram A quadrilateral having two pairs of parallel sides.

Pattern A model used in forming a mold for a casting.

Pentagon A five-sided polygon.

Percent(%) A fraction that indicates a part of 100.

Perimeter The distance around a two-dimensional figure.

Perpendicular Two or more lines or planes that intersect to form right angles.

Perspective drawing A pictorial drawing that contains receding lines that converge at a vanishing point on the horizon.

Pitch The distance from one point on a screw thread to the corresponding point on the next thread.

Plane A two-dimensional flat surface made up of points and lines.

Plot plan A map or drawing of an area that shows boundaries of lots and other parcels of property.

Polygon A closed figure in a plane formed by the intersection of three or more line segments at their endpoints. The number of angles in a polygon equals the number of segments.

Polyhedron A three-dimensional object having six or more plane faces.

Prism A six-sided polyhedron having two parallel faces called bases and whose other four faces are parallelograms.

Protractor A device used to measure or draw angles in degrees.

Quadrilateral A four-sided polygon.

Radius A line segment whose endpoints are the center of a circle and a point on the circle.

Ratio A fraction that compares to numbers or quantites.

Ream To enlarge a hole slightly to give it greater accuracy.

Rectangle A parallelogram having four right angles.

Reciprocal A number of the form $1/n$, where n is a non-zero number. The product of a non-zero number n and its reciprocal $1/n$ equals 1.

Rectangular prism A prism whose opposite faces are rectangles.

Regular polygon A polygon having equal sides and equal angles.

Rhombus A quadrilateral having four equal sides.

Right angle An angle whose measure is 90°.

Rivet A small fastener used to join objects together.

Round The outside rounded corner on a casting.

Scale A measuring device used to lay out or measure distances. In drafting, a factor used to enlarge or reduce the size of an object.

Scalene triangle A triangle having three unequal sides.

Schematic A diagram usually containing electric or electronic circuitry and symbols.

Sectional view A view of an object cut in half or quartered showing its interior detail.

Shaft A revolving bar, usually cylindrical, serving to transmit motion.

Shim A thin piece of material used as a spacer in adjusting two parts.

Sine (sin) In a right triangle, the ratio of the length of the leg opposite an acute angle to the length of the hypotenuse.

Spotface To produce a shallow circular bearing surface beneath the surface of a part.

Square A rectangle having four equal sides.

Straight angle An angle having a measure of 180°.

Structural drafter A person who prepares drawings dealing with the structural portion of buildings.

Surface area The sum of the areas of all the flat and/or curved surfaces of a three-dimensional object.

Surveyor A person skilled in land measurement.

Symmetry Correspondence of form and arrangement of parts on opposite sides of a boundary or axis, called a *line of symmetry* or *mirror line*.

Tangent (tan) A line perpendicular to a radius of a circle at its endpoint on the circle. In a right triangle, the ratio of the length of the leg opposite an acute angle to the length of the leg adjacent to the angle

Tap The tool used to cut internal threads.

Tap drill The drill used to make a hole for an internal thread.

Template A flat form that is used as a pattern or guide.

Tolerance A permissible variance in the dimension of a part.

Trapezoid A quadrilateral having only one pair of parallel sides.

Tread The step or horizontal member of a stair.

Triangle A three-sided polygon.

Triangulation A method of development using a system of imaginary triangles.

Truncate To cut off a geometric solid at an angle to its base.

Vellum Semitranslucent drawing paper used to make whiteprints.

Vernier scale A graduated scale used to obtain very precise measurements.

Volume A number that is a measure of the interior space of a three-dimensional object.

Weld Uniting metal parts by pressure or fusion welding processes.

Whiteprint A reproduction of a drawing made up of bluish lines on a light-colored background.

Working drawing A drawing containing the necessary information that will allow someone to work directly from it.

ANSWERS
TO ODD-NUMBERED PROBLEMS

Section 1
Whole Numbers

Unit 1 Addition

1. 220 in.
3. 841 mm
5. 221
7. 122
9. 37
11. 47
13. 107
15. a = 74,999
    b = 763,029
    c = 919,328
17. 11,767
19. 71
21. 562
23. 122
25. A = 229
    B = 66
27. 217
29. a = 46
    b = 30
31. A = 83
    B = 92
    C = 99
    D = 35
    E = 59

Unit 2 Subtraction

1. 92 ft.
3. 251 mm
5. 209 yd.
7. 402 ft.
9. 139
11. 33
13. 187
15. 34
17. 31
19. 28
21. 36
23. A = 24
    B = 27
    C = 10
25. 156
27. 5
29. 8

Unit 3 Multiplication

1. 666 cm
3. 5175 in.
5. 221 ft.
7. 864 yd.
9. 24
11. 119
13. a = $101.27
    b = $1,215.24
    c = $1,316.51
15. 150
17. 681

19. 3,050
23. 250
25. 282
27. A = 174
    B = 48
    C = 156
    D = 186
    E = 42
29. A = 315
    B = 162
    C = 207
    D = 405
    E = 612

Unit 4 Division

1. 83 in.
3. 63 cm
5. 38 mm
7. 6
9. 93
11. $4.00
13. a = 720
    b = 12
15. 4,789
17. a = 9 mm
    b = 11 mm
19. 15
21. No. of Windows = 22
    No. of Doors = 28
23. 60°
25. 37
27. 35
29. 17

Unit 5 Combined Operations

1. a = 1,269
   b = 1,221
3. 2,440
5. 24
7. 2,105
9. 54,398
11. 84 mm

13. A = 3 in.
    B = 2 in.
15. A = 35
    B = 44
    C = 61
17. 28
19. A = 5
    B = 8
    C = 22
    D = 31
21. Height = 48
    Width = 72
    Depth = 64
23. 28
25. A = 225
    B = 69
    C = 169
    D = 33
    E = 44
27. a = 19
    b = 12
    c = 114
    d = 13
    e = 59

Section 2
Fractions

Unit 6 Addition

1. 1 in.
3. ⅞ in.
5. 10 1/16 in.
7. 8⅔ hr.
9. 12 5/12 yd.
11. 8¾
13. 12⅝ in.
15. 3 5/16
17. A = 3½
    B = 3 13/16
19. A = 4 19/64
    B = 7 21/64

21. $7\frac{5}{32}$
23. $7\frac{5}{32}$
25. Height = 9 in.
    Width = 11 in.
27. $6\frac{16}{32}$
29. A = $2\frac{1}{2}$
    B = 3
    Height = $2\frac{1}{4}$
    Width = $5\frac{1}{4}$

Unit 7 Subtraction

1. $\frac{3}{8}$ in.
3. $1\frac{9}{16}$ yd.
5. $\frac{45}{64}$ in.
7. $1\frac{3}{4}$ lb.
9. $1\frac{5}{8}$ ft.
11. $\frac{15}{64}$
13. $18\frac{1}{4}$
15. $1\frac{3}{4}$ ft.
17. $9\frac{1}{2}$
19. C = $\frac{7}{16}$
    D = $\frac{23}{64}$
21. A = $3\frac{29}{32}$
    B = $4\frac{13}{32}$
23. $4\frac{35}{64}$ in.
25. A = $1\frac{9}{16}$
    B = $\frac{7}{16}$
27. A = $1\frac{49}{64}$
    B = $1\frac{1}{16}$
29. A = $2\frac{51}{64}$
    B = $2\frac{1}{16}$
    C = $\frac{31}{32}$
31. A = $3\frac{1}{16}$
    B = $2\frac{1}{8}$
    C = $\frac{3}{8}$
    D = $2\frac{9}{32}$
    E = $4\frac{15}{32}$

Unit 8 Multiplication

1. $\frac{3}{16}$
3. $\frac{7}{64}$
5. $\frac{13}{24}$

7. $17\frac{1}{2}$
9. $3\frac{33}{64}$
11. 42 in.
13. 675 cents or $6.75
15. $76\frac{7}{8}$
17. 21
19. $11\frac{1}{4}$
21. $5\frac{1}{4}$
23. Distance between floors = $101\frac{3}{4}$
    Dimension A = 132
25. a = $6\frac{1}{8}$
    b = $5\frac{1}{4}$
    c = $29\frac{3}{4}$
27. Height = 20
    Width = 36
    Depth = 10
    Rod diameter = 4
    Rod length = 26

Unit 9 Division

1. 4 in.
3. $20\frac{2}{5}$ yd.
5. $6\frac{1}{2}$ in.
7. $\frac{9}{98}$
9. $\frac{3}{7}$
11. $1\frac{27}{64}$
13. 128
15. 15
17. a = $25\frac{1}{2}$
    b = $27\frac{1}{4}$
19. $1\frac{1}{32}$
21. 160
23. $\frac{5}{8}$
25. Height = $\frac{3}{8}$
    Width = $1\frac{11}{16}$
    Depth = $1\frac{3}{8}$
27. A = $\frac{11}{32}$
    B = $\frac{3}{16}$
    C = $\frac{7}{16}$
    D = $\frac{9}{32}$
29. A = $2\frac{1}{16}$
    B = $2\frac{3}{38}$

Unit 10 Combined Operations

1. $2\frac{3}{128}$ in.
3. $10\frac{23}{24}$ lbs
5. 5
7. $1\frac{37}{64}$
9. 80
11. 39
13. $1\frac{1}{2}$
15. $2\frac{3}{8}$
17. $5\frac{1}{2}$
19. $1\frac{1}{4}$
21. Cabin perimeter = 85 ft. 0 in. or 85′ 0″
    Deck perimeter = 42 ft. 0 in. or 42′ 0″
    Difference = 43 ft. 0 in. or 43′ 0″
23. Height = $8\frac{7}{8}$ in.
    Width = $10\frac{15}{16}$ in.
    A = $2\frac{7}{8}$ in.
    B = $3\frac{1}{2}$ in.
25. A = $1\frac{17}{32}$
    B = $5\frac{57}{64}$
    C = $2\frac{1}{32}$
    D = $2\frac{7}{16}$
    E = $2\frac{1}{8}$
27. A = 4 ft. 6 in. or 4′ 6″
    B = 3 ft. 6 in. or 3′ 6″
    C = 8 ft. 6 in. or 8′ 6″
    D = 7 ft. 0 in. or 7′ 0″
    Perimeter = 78 ft. 0 in. or 78′ 0″
29. A = $\frac{57}{64}$
    Diameter (1) = $3\frac{1}{8}$
    Diameter (2) = $13\frac{3}{4}$
31. A = $8\frac{7}{8}$
    B = $\frac{15}{16}$
    C = $4\frac{11}{16}$
    D = $4\frac{13}{16}$
    E = $1\frac{9}{32}$

Section 3
Decimals

Unit 11 Addition

1. 0.887
3. 0.8772
5. 1.2605
7. 45.346
9. $53.41
11. 40.75
13. 23.57
15. 429.05
17. 200.92
19. 135.03
21. 3.62
23. 366.64
25. A = 6.008
    B = 3.009
    C = 4.088
27. A = 4.79
    B = 1.07
    C = 2.38
29. A = 6.42
    B = 4.63
31. Inside perimeter = 10
    Outside perimeter = 36.5

Unit 12 Subtraction

1. 0.59
3. 1.066
5. 7.71
7. 40.817
9. 1.6558
11. $169.15
13. $150.99
15. 0.0157
17. $1,680.95
19. 2.336
21. A = 0.4319
    B = 0.8946

23. A = 0.078
    B = 1.142
    C = 0.237
25. 8
27. 154.6
29. A = 2.25
    B = 0.75
    C = 1.69
    D = 0.37
    E = 0.38
31. A = 242.1
    B = 94.5
    C = 48.1
    D = 86.6

Unit 13 Multiplication

1. 0.0498
3. 0.40455
5. 3.6828
7. 22.2042
9. 3
11. $648.00
13. $759.36
15. 14.4375
17. $119.70
19. 2.4375
21. 9.30
23. 139.50
25. Height = 3.75
    Width = 3.75
27. Height = 1.75
    Width = 2.625
29. A = 0.563
    Width = 8.764
    Height = 5.264

Unit 14 Division

1. 0.24
3. 7.0
5. 0.0003

7. 29.31
9. 24
11. 0.036
13. $2,136.56
15. 37
17. $1.87
19. 2.757
21. 30.20
23. 41.6715
25. Height = 0.50
    Width = 0.375
27. A = 0.54
    B = 1.06
    C = 0.16
    D = 0.343
    E = 0.600

Unit 15 Combined Operations

1. 0.176
3. 12.199
5. 273.058
7. 1.4469
9. 5.757
11. a = $1,142.55
    b = 1,116
13. a = $4,148.98
    b = $3,848.99
    c = $20,744.90
15. A = 7.190
    B = 1.221
17. Match plate length = 13.27
    Dimension A = 0.505
19. A = 1.85
    B = 1.45
21. Height = 197.5
    Width = 241.2
23. Height = 4.267
    Width = 7.438
25. Height = 5.560
    Width = 12.275

Section 4
Decimals, Fractions, and Percents

Unit 16 Equivalent Decimals and Fractions

Exact and approximate answers given for some problems.

1. 0.286
3. 0.458
5. $\frac{3}{16}$
7. $\frac{13}{16}$
9. 2.313 in.
11. $28\frac{3}{4}$
13. a = $7\frac{15}{32}$
    b = 7.469
15. $2\dfrac{5156}{10,000} \approx 2\frac{33}{64}$
17. A = $1\dfrac{187}{1000} \approx 1\frac{3}{16}$
    B = $\dfrac{203}{500} \approx \frac{13}{32}$
19. $\frac{3}{32}$
21. $1\frac{7}{8}$
23. A = 0.656
    F = 5.984
25. A = 0.625
    B = 2.313
27. $\frac{17}{32}$

Unit 17 Percents

1. 3.36
3. 15
5. 250.32
7. 9 unit-squares are shaded
9. 18
11. 101.37
13. 17
15. 345
17. 27
19. Total layers used = 78%
    Plan views = 38%
    Elevations = 25%

Details = 20%
Construction = 13%
Miscellaneous = 5%
21. $91.82
23. a = $1,150.00
    b = $287.50
    c = $99.45
    d = $763.15
25. A = 1.493
    B = 5.206
27. 3.402

Unit 18 Simple Interest and Discounts

1. $4,320.00
3. 13%
5. a = $0.07
    b = $195.30
7. $2.63
9. $804.88
13. a = $267.53
    b = $414.24
    c = $181.23
15. a = $513.60
    b = $433.89
    c = $18.08
17. $866,400
19. $157.50
21. $1,368.00
23. $12,470.00
25. $15\frac{1}{4}$%
27. $558.75
29. $1,014.00

Section 5
Geometry

Unit 19 Powers, Roots, and the Pythagorean Theorem

1. 9
3. 81
5. 64
7. 2,197

9. 15.625
11. 0.421875
13. 7
15. 2.61
17. 1.378
19. 1.29
21. 14.75
23. 8.27
25. 18
27. 4,330.747
29. 6
31. 7.60
33. 6.5

Unit 20 Lines, Angles, and Triangles

1. Obtuse
3. 90°
5. Triangle A = equilateral and acute
   Triangle B = isosceles
   Triangle C = right
7. 56°
9. Right
11. a = Scalene
    b = 6.25
    c = ∠B

Unit 21 Polygons and Circles

1. 48
3. 27
5. 80
7. 33.6
9. 15.70
11. 21.6
13. 50.27

Unit 22 Constructions

You can find possible solutions for the odd-numbered constructions, 1 to 19, in the Instructors Guide. There may be other possible solutions as well. See your Instructor to check your construction for problem 21.

Section 6
Measurement

Unit 23 Linear Measure

1. a = 84 in.
   b = 47 in.
   c = 67¾ in.
   d = 75¼ in.
3. a = 45 in.
   b = 59.25 in.
   c = 64.44 in.
   d = 140.38 in.
5. A = $\frac{7}{64}$ in.
   B = $\frac{35}{64}$ in.
   C = $\frac{61}{64}$ in.
   D = $1\frac{13}{64}$ in.
   E = $\frac{9}{32}$ in.
   F = $\frac{25}{32}$ in.
7. A = 12 mm
   B = 23 mm
   C = 36 mm
   D = 47 mm
   E = 59 mm
   F = 74 mm
9. a = 3.376 in.
   b = 2.641 in.
   c = 2.021 in.
11. A = 0′−2″
    B = 1′−4″
    C = 1′−8″
    D = 2′−10″
13. a = 16′−3″
    b = 18′−9″
    c = 19′−6″
    d = 7′−6″
    e = 11′−0″
    f = 9′−0″
    g = 9′−10½″
    h = 8′−9″
    i = 1′−10″
    j = 2′−2″

k = 5′ − 2″
l = 1′ − 8¾″
m = 1′ − 1⅞″
n = 0′ − 3½″

15. Deck = 144′ 0″
Bedroom = 65′ 6″
Living room = 73′ 0″

Unit 24 Scaled Measurements

1. Line a = 4¾ in.
   Line b = 3¼ in.
3. Line a = 2¹¹⁄₁₆ in.
   Line b = 4⁹⁄₁₆ in.
5. Line a = 1⁴³⁄₆₄ in.
   Line b = 3⁵³⁄₆₄ in.
7. Line a = 4²⁵⁄₃₂ in.
   Line b = 2⅜ in.
9. A = 2¾ in.
   B = 4½ in.
   C = 2⅛ in.
   D = 8¼ in.
   E = 17½ in.
   F = 5½ in.
11. A = 18 mm
    B = 16 mm
    C = 25 mm
    D = 28 mm
    E = 83 mm
    F = 39 mm
    G = 15 mm
    H = 11 mm
13. A = ¾ in.
    B = ⅜ in.
    C = 2¾ in.
    D = 3¾ in.
    E = 1⅝ in.
    F = 1¼ in.
15. A = 288′
    B = 290′
    C = 114′

D = 305′
E = 236′
F = 99′
G = 145′

Unit 25 Area

1. 216
3. 56.70
5. 7,169
7. 10.563
9. 7,290
11. 9.094
13. 177.48
15. 15 lb.
17. 9.695
19. 306.51
21. Area of cabin = 472.875 ft.^{2}
    Area of deck = 109.25 ft.^{2}
23. 188
25. A = 61,600
    B = 53,200
    C = 30,000
    D = 40,000
27. Total area = 4,048,750
    Lot 1 = 756,250
    Lot 2 = 900,000
    Lot 3 = 550,000
    Lot 4 = 1,065,000
    Lot 5 = 777,500
29. 63.617
31. 7.915
33. 135.89
35. 78.93
37. 197.921
39. 114.670
41. 88.439
43. 138.090
45. Area of template = 20.824
    % waste = 26
47. 1.7188

49. 24

51. 3,018.84

Unit 26 Volume

1. 10,368
3. 175
5. 2,114
7. 784.21
9. 3.35
11. 298.67
13. ¼
15. 3,990
17. 81.844
19. 788
21. 66.366 in.^{3}
23. 21.21
25. 753.98
27. 2,865.139 mm^3
29. 92.78
31. 2.105
33. 4,008.76
35. 39.58
37. 1.687
39. 53.041
41. 13.2014
43. 15.70
45. 40.98
47. 7.76

Unit 27 Equivalent Measurement Units

1. 228.600 mm
3. 41.91 cm
5. 5.791 m
7. 156.162 in.
9. 66.929 in.
11. 406.4
13. 17
15. 14.405
17. 6.6

19. a = 0.588
    b = 2.477
21. 109.68
23. 294.5
25. 17.8142
27. 78.54
29. 71.77
31. 39.773
33. 0.363
35. 3.575
37. 215.99
39. 201.830
41. 69.59
43. 13.76
45. 939.75
47. 1,647.62
49. 200.04
51. 1,394.29
53. 83.35
55. 41.50
57. 10.90
59. 38.28
61. 41.44

Unit 28 Angle Measure

1. 180°
3. 360°
5. 26° 10′
7. 57° 45′
9. 18° 18′
11. 15°
13. 56° 43′
15. 114° 33′
17. 45° 27′
19. 72°
21. ∠ A = 90°
    ∠ B = 90°
    ∠ C = 15°
    ∠ D = 120°
    ∠ E = 30°
    ∠ F = 90°

23. $\angle A = 105°$
     $\angle B = 143°$
     $\angle C = 120°$
     $\angle D = 90°$
25. $\angle C = 78° \ 58'$
     $\angle D = 22° \ 42'$

Section 7
Algebra

Unit 29 Expressions and Equations

1. 6
3. 1.04
5. 35
7. $x = 46$
9. $Z = 8\frac{1}{2}$
11. 57H
13. 90
15. ¼
17. $A = 7\frac{1}{8}L$
     $B = 6\frac{13}{16}L$
19. $X = 0.5$
     $A = 1.0$
     $B = 2$
     $C = 1.8$
     $D = 1.5$
21. 8.85

Unit 30 Ratios

1. ³⁄₂
3. 8:1
5. $^{48}\!/_{12} = ^4\!/_1$
7. ¹⁄₁₆
9. 4:1
11. 5 times faster
13. 1:9
15. 1:48
17. 1:8

19. $A = 3^{15}\!/_{32}$
     $B = 7^5\!/_{16}$
     $C = 1^{19}\!/_{32}$
21. $A = 1\frac{1}{16}$
     $B = {}^{13}\!/_{16}$
     $C = {}^7\!/_{16}$
     $D = 1^{11}\!/_{32}$
23. 1:5

Unit 31 Proportions

1. $x = 15$
3. $Y = 15.25$
5. 15
7. 36
9. $x = 152$ min.
11. $x = 25$
13. $x = 26$
15. $x = 4.44$ in.
17. $a = 12$
     $b = 72$
19. 15
21. 2 min. 42 sec.
23. 15 min.

Unit 32 Formulas and Handbook Data

1. 177.80
3. 76.20
5. 0.62
7. 7.57
9. 21.99
11. ¹⁵⁄₁₆
13. 160°
15. 75.65
17. 0.500
19. 21.206
21. $a = 6.5$
     $b = 0.25$
     $c = 0.393$
     $d = 0.539$

Unit 33 Averages

1. $26.50
3. 37.20 mi.
5. 83
7. $^{37}/_{64}$
9. $560.33
11. 15
13. 9.7
15. 39
17. Avg. height = 0.439
    Avg. width = 0.612
    Avg. hole diam. = 0.453

Section 8
Graphing

Unit 34 Coordinate Systems

1. 126
3. Height = 16
    Width = 24
5. Height = 2.5
    Width = 1.75
7. Height = 35
    Width = 49
    Diameter = 14
9. Height = 3¹⁄₁₆
    Width = 2³⁄₁₆
11. Height = 18.38
    Width = 15.75

Unit 35 Statistical Graphs

1. a = Labor
    b = Supplies
    c = 24%
3. a = 2010
    b = 2006
    c = 2009
    d = 1,250

5. a = 900 hours
    b = October
    c = April
    d = October, November, and December

7.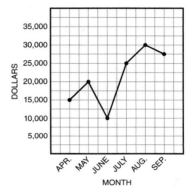

Section 9
Applied Trigonometry

Unit 36 Right Triangles

1. 0.97029
3. 0.57358
5. 55°
7. cos 61°
9. sin 36°
11. 0.88928
13. 77°
15. 22 in.
17. 7
19. AC = 2.483
    BC = 1.049
21. 6.93
23. 7.7024
25. 0.488
27. 3.325
29. A = 5.546
    B = 3.061

31. 3.80
33. Line A = 4.07
    Line C = 1.48
35. a = 4.696
    b = 41°52′
37. Line A = 4.89
    Line B = 3.70
39. Shorter leg = 5.77 in.
    Hypotenuse = 11.54 in.

Unit 37 Oblique Triangles

1. 0.89662
3. 52° 55′ 50″
5. 52° 16′ 41″
7. −0.77521
9. 1.749
11. a = 2.016
    ∠ C = 33° 50′
13. ∠ A = 65° 8′
    ∠ B = 93° 35′
    ∠ C = 21° 17′
15. 73° 31′
17. 4.62
19. 0.382
21. Line A = 3.536
    Line B = 4.106
    Line C = 2.790
23. 3.86
25. Line AB = 1.69
    Line AD = 1.15

Section 10
Estimation and Tolerance

Unit 38 Estimation and Percent Accuracy

1. 86.8%
3. About 74% of estimate or about 26% below
   the estimate.

5. About 94% of estimate or about 6% below
   the estimate.
7. 4%
9. About 134% of estimate or about 34% above
   the estimate
11. About 97% of estimate or about 3% below
    the estimate

Unit 39 Tolerance and GD&T

1. .034
3. Upper limit = 4.390
   Lower limit = 4.366
   Tolerance = 0.024
5. 2.587
7. 4.392
9. Hole tolerance = 0.0012
   Shaft tolerance = 0.0012
11. 0.14005
13. $\dfrac{1.490}{1.485}$

15. A = 0.125
    B = ¹⁄₁₆
17. Upper Limit = $\dfrac{1.373}{1.379}$

    Lower Limit = $\dfrac{1.372}{1.368}$

    Allowance = 0.001
    Type of fit = Clearance
19. Allowance = −0.0022
    Hole tolerance = 0.0010
    Shaft tolerance = 0.0009
21. A = 0.444
    B = 2.357
    C = 0.658

9